KENDALL/HUNT PUBLISHING COMPANY
4050 Westmark Drive Dubuque, Iowa 52002

Book Team
Chairman and Chief Executive Officer Mark C. Falb
Vice President, Production Services Al Grisanti
Director of National Book Program Paul B. Carty
Editorial Development Manager Georgia Botsford
Developmental Editor Tina Bower
Assistant Vice President, Production Services Christine E. O'Brien
Prepress Project Coordinator Angela Shaffer
Designer Suzanne Millius

Front Cover Photograph:
Spring wildflowers at mouth of Natural Bridge Canyon in the Black Mountains. The hills in the foreground consist mostly of Pleistocene alluvial fan deposits; the mountain in the background consists of Late Tertiary volcanic rock, and the slope on the right is underlain mostly by Precambrian metamorphic rocks. The Badwater turtleback fault separates the metamorphic rocks from the volcanic rocks and sediments.

Title Page Photograph:
Geologist near base of Copper Canyon Turtleback in the southern Black Mountains. Smith Mountain and Mormon Point lie in the background.

Back Cover Photographs:
Bottom photo: Moonrise over volcanic rock, Salisbury Pass.
Middle photo: Saline Valley from near Racetrack Playa.
Top photo: Recumbent Syncline at Corkscrew Peak in the Grapevine Mountains.

ISBN 0-7575-0950-9

Printed in the United States of America
10 9 8 7 6 5 4 3 2 1

Contents

Chapter Four Death Valley Road Guides 63

Appendices 109

About the Authors

Marli Miller has studied and photographed geology in Death Valley since she was introduced to the area by Lauren Wright in 1984. Much of her research focuses on the evolution of the Black Mountains turtlebacks, although she spends more and more of her time in the Amargosa Chaos. She completed her B.A. in geology at Colorado College in 1982 and then her M.S. and Ph.D. at the University of Washington in 1987 and 1992 respectively. Marli lives with her partner and two daughters in Eugene, Oregon. There, she is a senior instructor and researcher at the University of Oregon.

Lauren Wright began his geological investigations of the Death Valley region in the late 1940's while working toward his Ph.D. degree at Caltech. He and his field partner Bennie Troxel were both employed by the California Division of Mines and Geology and cooperated in the preparation of geologic maps of much of the southern part of this region. In 1961 Wright moved eastward to teach geology at Pennsylvania State University. He continued his association with Troxel and supervised the Death Valley-oriented research of numerous graduate students. Since his retirement in 1985 he returns to Death Valley almost yearly to work in cooperation with Troxel, a number of his former students and other younger geologists. He observes that his association with Marli Miller has been especially gratifying and rewarding.

Preface

In a geologic sense, few places in North America can rival Death Valley. Death Valley's intense aridity means that the bedrock and landforms of the region are exceptionally well-exposed. It also hosts a diverse assemblage of rock types that formed over an unusually long span of time. Finally, Death Valley's location is a focal point for North American geology: the region shows the effects of nearly all the principal geologic events that together shaped western North America. These events include the Late Proterozoic rifting of western North America, intrusion of vast quantities of granitic igneous rock during the Mesozoic Era, and crustal extension during the late Cenozoic Era and continuing today. For this latter event, Death Valley is truly a world-class locality.

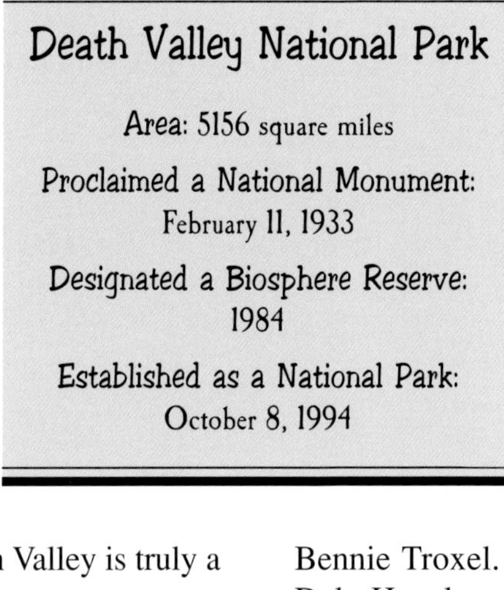

Death Valley National Park

Area: 5156 square miles

Proclaimed a National Monument: February 11, 1933

Designated a Biosphere Reserve: 1984

Established as a National Park: October 8, 1994

In keeping with the visibility of Death Valley's geology, this book employs numerous photographs as well as text to introduce the geology. It presents the geologic features in three chapters. The first chapter describes many of the landforms and geologic processes that are so visible in Death Valley. The second highlights late Cenozoic extension and how it has shaped, and is still shaping, the region. The third outlines the region's geologic history.

This book began as a greatly revised and expanded version of the chapter on Death Valley in the book *Geology of National Parks*, edited by Harris, Tuttle and Tuttle (Wright and Miller, 1997). This second edition is further updated and modified to include guides to all the major roads in Death Valley. We benefited greatly from discussions with numerous people on the region's geology, most notably Darrel Cowan, Terry Pavlis, Mitch Reynolds, Laura Serpa, and Bennie Troxel. Kathy Cashman, Darrel Cowan, Dale Housley, and Mark Neuweld critiqued an early version of the manuscript, and Mel Essington gave valuable help on the short mining section. We also want to thank Charles Rogers and Craig Smith, who piloted the planes from which Miller shot the aerial photographs.

Introduction

Geographic Setting, Fault Zones, and Rocks

True to its reputation, the floor of Death Valley is, indeed, the hottest, driest, and lowest land in the United States. Summer temperatures frequently exceed 49°C (120°F) and the mean annual precipitation is only about 4.8 cm (1.9 inches). But, more importantly, Death Valley National Park encloses one of the world's most spectacular desert landscapes and a mountain range that remains snow-capped during most winter seasons (Photo 1). Even on the floor of the valley temperatures occasionally drop below freezing.

The valley lies in the Mojave Desert of southeastern California and is arid because it lies in the **rain shadow** of several more westerly mountain ranges including the Sierra Nevada (Fig. 1). It also contains the low point of a large region of interior drainage in the southwestern part of the Great

Rain Shadow. An area on the downwind side of a mountain range. Rain shadows are especially arid because the winds lose their moisture as they rise over the mountains.

Photo 1. View northwestward into central Death Valley. The Panamint Range lies on the left (west), the Black Mountains on the right, and the Grapevine Mountains in the background. Mormon Point lies in the left foreground. Mormon Point appears to protrude into the valley because it was uplifted along several different fault zones (Pavlis et al., 1993). These faults lie at the valley floor/mountain front transition.

Basics About...

Faults and Fault Zones

Faults are simply fractures along which the adjacent rock has moved. Faults are classified according to how the rock has moved with respect to the other side.

Dip slip faults are those in which the rock moved up or down the "dip" (tilt direction) of the fault plane. They are further classified as either "normal" or "thrust" faults. To determine if a fault is normal or thrust, it is first necessary to distinguish the hangingwall (the block of rock that rests on top the fault) from the footwall (the block of rock that rests below the fault). In the accompanying photographs, the hangingwall is labelled "HW"; the footwall is labelled "FW."

Photo 1-1. Normal fault behind Artist Drive. Note how the layers tilt in the direction opposite the dip of the fault.

Normal faults are those in which the hangingwall moved down with respect to the footwall. Normal faults typically place younger rock on top of older rock. They form by extension of the earth's crust and are the predominant fault in Death Valley. See Photo 1-1.

Photo 1-2. Thrust fault and folded rock.

Thrust faults are those in which the hangingwall moved up with respect to the footwall. Thrust faults typically place older rock on top of younger rock. They form by compression of the earth's crust, and so are typically accompanied by folded rock. In Death Valley, most thrust faults formed during the Mesozoic Era. Thrust faults are typically inclined at 40° or less; those that are steeper are called "Reverse" faults. See Photo 1-2.

Strike-slip faults are those in which the rock moved along the "strike" (horizontal length) of the fault plane. They are further classified as either "right-lateral" or "left-lateral" depending on the relative sense of offset of the faulted rock.

Photo 1-3. Right-lateral strike-slip fault.

Right-lateral faults are those in which the block opposite the viewer moved to the right. Notice that it does not matter which side of the fault the viewer is standing. In Death Valley, northwest-trending right-lateral faults are the dominant strike-slip fault type. See Photo 1-3.

Left-lateral faults are those in which the block opposite the viewer moved to the left.

Oblique-slip faults are those in which the rock moved neither parallel to strike nor dip, but in a sense that includes components of each. See Diagram 1-4.

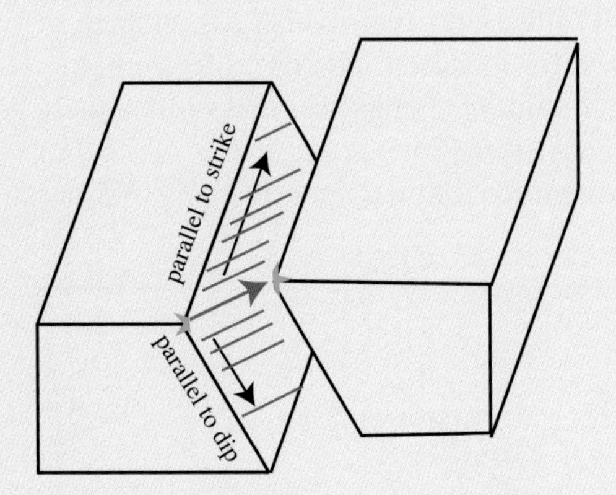

Diagram 1-4. Block diagram of an oblique-slip fault. Black arrows show direction of dip and strike; red arrow shows actual slip direction on fault with components of both dip and strike slip. Note that restoring the fault in a direction parallel to the red lines restores the green star to its original shape.

Figure 1. Landsat image of Death Valley region with park boundaries shown in white. Modified from Thelin and Pike, 1991.

Basin. The valley floor, which extends 86 m (282 feet) below sea level near Badwater, lies between north- to northwest-trending mountain ranges. The Panamint Mountains, west of central Death Valley, culminate in Telescope Peak at an elevation of 3368 m (11,049 feet). The relief between there and the valley floor is one of the greatest obtained within the conterminous United States.

This relief results from the phenomenon of crustal extension, which throughout the entire region has caused mountain ranges to rise and intervening valleys to sink along fault zones. Most of the active **faults** are normal faults (see Basics About Faults and Fault Zones, pp. viii, ix). Besides allowing the crust to extend, normal faults can also cause rocks to tilt in a direction opposite the inclination, or dip, of the fault. Consequently, the Death Valley region is marked by numerous "tilted fault blocks"—block-like mountain ranges that have risen and tilted along normal fault zones (Fig. 2).

Faults. Breaks or fractures in rock along which one side has moved relative to the other. The type of relative movement determines the type of fault (see Basics About Faults and Fault Zones).

Some striking examples of tilted fault block ranges within the national park are the Last Chance Range and the Panamint, Cottonwood, and Black Mountains. The Resting Spring and Nopah Ranges, east of the park boundary, are also tilted fault blocks. Each range trends approximately northward, is bound by a west-dipping fault, and tilts eastward. This apparent simplicity, however, is deceptive as each of these ranges contains other extension-related, and even older, faults. Chapter two treats extension-related faulting in much greater detail.

The geologic map (Fig. 3A, 3B) illustrates the diverse assemblage of rock types that make up Death Valley. Basics About Igneous, Metamorphic, and Sedimentary Rocks (pp. xiv–xv) outlines the processes that formed these rocks. Most of the present ranges consist of Proterozoic, Paleozoic and/or Mesozoic rocks that pre-date the formation of these ranges by many millions

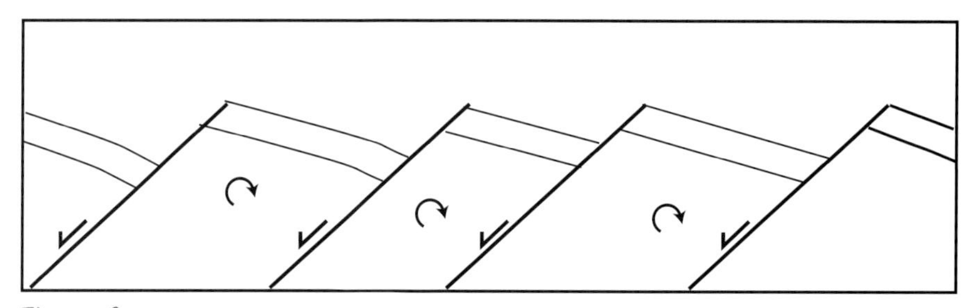

Figure 2. Fault-block ranges. As slip takes place along large normal faults, the rocks become tilted in a direction that is opposite to the dip on the fault.

Death Valley National Park, California

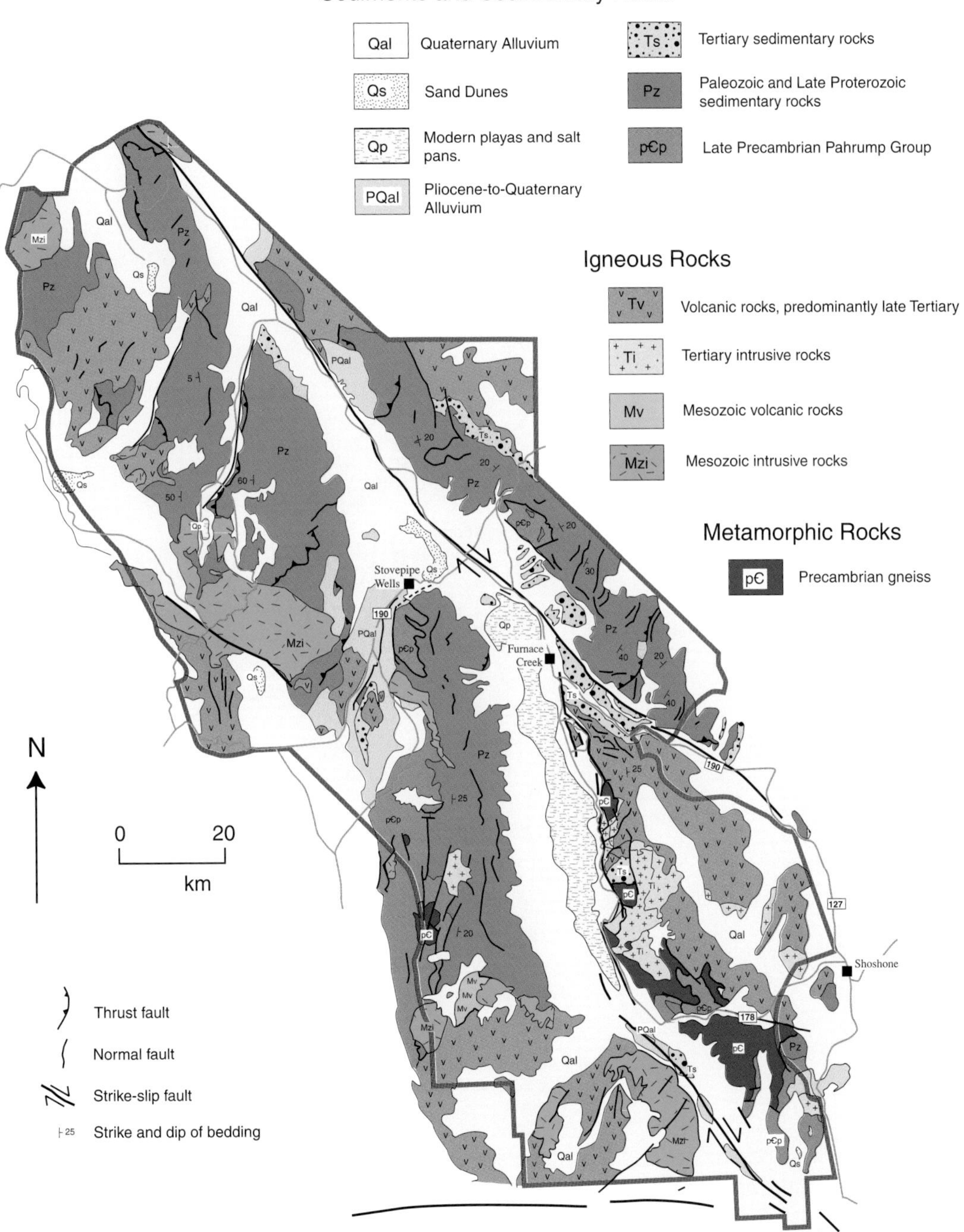

Figure 3A. Geologic map of Death Valley National Park, compiled largely from Jennings et al., 1962, Strand, 1967, and Streitz and Stinson, 1974. Some specific data derived from Albee et al., 1981, Hall, 1971, Burchfiel, 1969, McAllister, 1956, 1970, Reynolds, 1969, Wright and Troxcl, 1984, 1993. Map image first published by Miller, 2001.

Death Valley National Park, California

Location Map

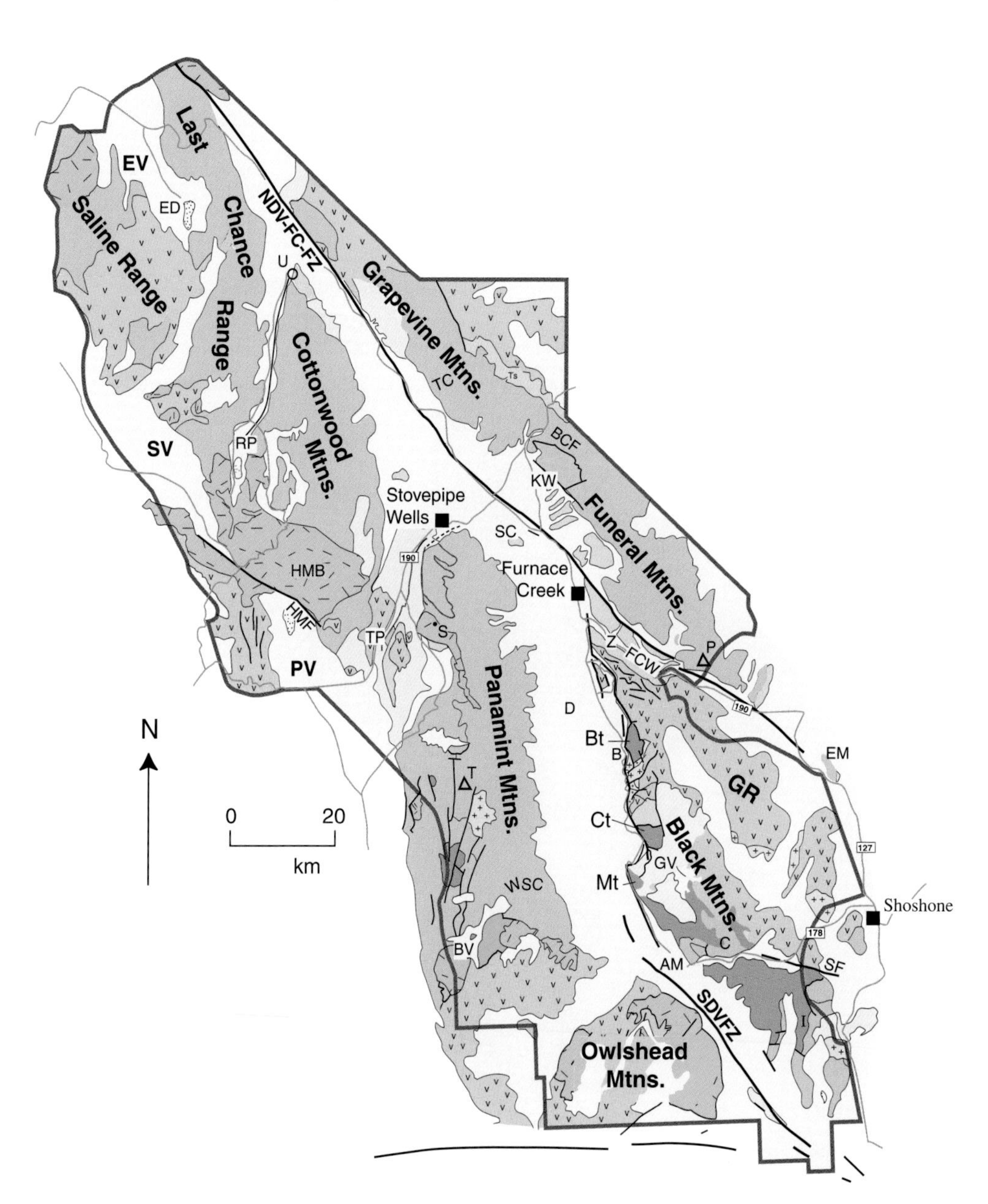

Figure 3B. Location map showing the principal mountain ranges of Death Valley National Park. Abbreviations for locations are as follows: AM, Ashford Millsite; B, Badwater; BCF, Boundary Canyon Fault; Bt, Badwater Turtleback; BV, Butte Valley; C, Chaos; CT, Copper Canyon Turtleback; D, Devil's Golf Course; ED, Eureka Dunes; EM, Eagle Mountain; EV, Eureka Valley; FCW, Furnace Creek Wash; GR, Greenwater Range; GV, Gold Valley; HMB, Hunter Mountain Batholith; HMF, Hunter Mountain Fault; I, Ibex Hills; KW, Keane Wonder Mine; Mt, Mormon Point Turtleback; NDV-FC-FZ, Northern Death Valley-Furnace Creek Fault Zone; P, Pyramid Peak; PV, Panamint Valley; RP, Racetrack Playa; S, Skidoo; SC, Salt Creek; SDVFZ, Southern Death Valley Fault Zone; SF, Sheephead Fault; SV, Saline Valley; T, Telescope Peak; TP, Towne Pass; U, Ubehebe Crater; WSC, Warm Springs Canyon; Z, Zabriskie Point.

of years (Table 1). In fact, we owe much of our knowledge of the earlier events to exposures of these rocks along the tilted faces of the ranges. We observe in the oldest rocks evidence of events that shaped and metamorphosed the crust more than 1.7 billion years ago. An even clearer record of ancient marine and river environments and accompanying igneous activity is contained in the extensive exposures of the Pahrump Group, emplaced within the 1.2 to 0.8 billion year interval of time. The stark mountain slopes also contain continuous exposures of the later Proterozoic and Paleozoic formations that record the pre-Mesozoic history of the western margin of North America.

The contiguous Black Mountains and Greenwater Range, in the eastern part of the park, are underlain largely by bodies of Tertiary igneous rocks, both extrusive and intrusive. These were added to the pre-existing crust 12 to 4 million years ago. When the Tertiary record is combined with that of the older rocks preserved in the ranges, geologists can reconstruct an extraordinarily complete record of crustal evolution. This record is detailed in Chapter 3.

Basics About...
Igneous, Metamorphic, and Sedimentary Rocks

All rocks fall into one of three groups: igneous, metamorphic, or sedimentary. Igneous and metamorphic rocks form within the earth at high temperatures. Sedimentary rocks form at the earth's surface.

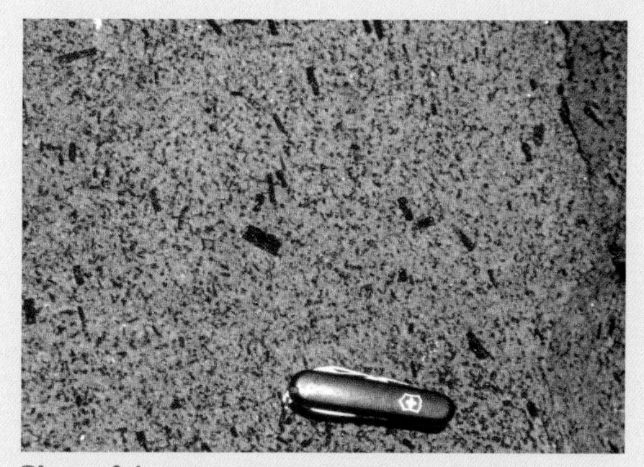

Photo 2-1. Intrusive igneous rock, granite, showing coarsely crystalline texture.

Igneous rocks form by cooling and crystallization from a molten state. Therefore, they consist of interlocking crystals. Intrusive igneous rocks cool and crystallize slowly within the earth and so have coarse grains. See Photo 2-1. Extrusive (volcanic) igneous rocks cool and crystallize rapidly on the earth's surface and so have fine grains. Extrusive rocks frequently contain some larger, well-shaped, crystals surrounded by the fine-grained material.

Metamorphic rocks consist of pre-existing (igneous, metamorphic, or sedimentary) rocks that underwent a change because of high temperatures and usually, pressure. With these high temperatures, the original mineral grains react and change into different types. Clay minerals, for example, will transform into micas. Because the metamorphism usually also involves directed pressures, the newly formed minerals grow in a preferred orientation. Consequently, metamorphic rocks have a crystalline texture and usually show a distinctive layering, called foliation. See Photo 2-2.

Photo 2-2. High grade metamorphic rock (gneiss).

One can distinguish foliation from bedding in sedimentary rocks because foliation is formed by oriented *crystals* whereas bedding is formed by different-sized, angular to rounded *grains*.

Metamorphic rocks are further characterized by their grain size. In general, the coarser the grain size, the higher "grade," or temperature, of metamorphism.

Sedimentary rocks consist of particles (sediment) worn from pre-existing rocks. They may be *clastic, chemical, or biogenic* in origin. Clastic sedimentary rocks consist of gravel, sand, silt, clay. Chemical sedimentary rocks consist of precipitated materials (eg. Salt). Biogenic sedimentary rocks consist of biologically produced material (eg. shell fragments). The most com-

Photo 2-3A. Limestone with visible crinoid fossils.

Photo 2-3B. Alternate beds of sandstone and shale.

mon biogenic sedimentary rocks, limestone and dolomite, are widespread in Death Valley (See Photo 2-3A). In all cases, the sediment accumulates in layers, and through time, becomes glued together through the action of groundwater. At the scale of an outcrop, one can see this layering, called "bedding" or "stratification" (See Photo 2-3B). Single hand specimens are usually too small to show bedding.

Clastic sedimentary rocks are further classified according to grain size. Those with the smallest (clay-sized) particles are called shale. Those with siltsized particles are called siltstone; those with sand-sized particles are called sandstone (See Photo 2-3C); those with pebble-sized and larger particles are called conglomerate (See Photo 2-3D).

Photo 2-3C. Close-up view of sandstone.

Photo 2-3D. Conglomerate.

Time Units			Rock Units		Principal Geologic Events
Era	Period	Epoch	Group	Formation	
Cenozoic	Quaternary	Holocene	Alluvial fans, stream and playa deposits, dunes		Continued deposition in modern Death Valley
	Tertiary	Pliocene Miocene	Numerous sedimentary, volcanic and plutonic units in separate and interconnected basins and igneous fields; includes Artist Drive, Furnace Creek, Funeral, and Nova Formations.		Opening of modern Death Valley Continuing development of the present ranges and basins Onset of major extension
				Several formations	Deposition on relatively subdued terrain
		Oligocene		Titus Canyon	
			— — — — — — — — — — — — Major	Unconformity — — — — — —	
Mesozoic	Cretaceous/ Jurassic		Granitic plutons		Thrust faulting and intrusion of plutons related to Sierra Nevada batholith
	Triassic			Butte Valley	Shallow marine deposition
				— — — — — —	Unconformity — — — — — —
Paleozoic	Pennsylvanian			Resting Spring Shale	Development of a long-continuing carbonate bank on a passive continental margin; numerous intervals of emergence, interrupted by deposition of a blanket of sandstone in Middle Ordovician time.
	Mississippian			Tin Mountain Limestone Lost Burro	
	Devonian/ Silurian			Hidden Valley Dolomite	
	Ordovician			Ely Springs Dolomite Eureka Quartzite	
			Pogonip		
	Cambrian			Nopah Bonanza King Carrara Zabriskie Quartzite Wood Canyon	Deposition of a wedge of siliciclastic sediment during and immediately following the rifting along a new continental margin
Prototerozoic				Stirling Quartzite Johnnie Ibex Noonday Dolomite	Shallow to deep marine deposition along an incipient continental margin
				— — — — — —	Unconformity — — — — — —
			Pahrump	Kingston Peak Beck Spring Crystal Spring	Glacio-marine deposition Shallow marine deposition Rapid uplift and erosion
			— — — — — — — — — — — — Major	Unconformity — — — — — —	
			Crystalline basement		Regional metamorphism

Table 1. Stratigraphic column of Death Valley.

Chapter One

Death Valley's Active Geology

Moonrise over sand dunes at Mesquite Flat.

Active Faulting Along the Black Mountains Front

Alluvial fans. Fan-shaped landforms, found typically at the bases of mountains, of material eroded from the mountains. Most strictly, this material consists of alluvium, which is water-transported, but in many cases it also consists of material that was transported primarily by gravity.

Fault scarp. A sudden rise or step in the landscape that formed by slip on a fault zone.

Transition from Valley Floor to Mountain Front

The extraordinarily abrupt topographic break between the Black Mountains escarpment and the sedimentary fill of the valley marks the approximate location of the Black Mountains fault zone (Photo 2). Here the mountains rise while the valley floor drops. In most places, relatively small **alluvial fans** spill out of deep canyons at the fault zone; where there are no canyons, the horizontal deposits of the valley floor meet a wall of rock. Much of the mountain front, therefore, qualifies as a **fault scarp**, an exposed surface produced directly by movement on a fault and essentially unmodified by erosion or weathering.

The morphology of the Black Mountains escarpment provides important information about the nature of the fault that produced it. When viewed from a distance, such as from the Devil's Golf Course, parts of the mountain front appear incredibly smooth and probably do closely coincide with the actual fault surface. One can see how the Black Mountains lie in the footwall of the fault, and so the relief is an effect of crustal extension.

Fault Scarps

Fault scarps in alluvial fans along the Black Mountains front provide additional and direct evidence that the front is presently active. That they are fault-generated rather than stream-cut is attested by the observation that they cut across stream-related features. Particularly obvious fault scarps cut the fan immediately south of Badwater (Photo 3) and other fans at Mormon Point. The scarp at Badwater is aligned with the spring that feeds the Badwater pond and suggests that the spring is controlled by the fault. Other scarps are well defined between Furnace Creek Inn and Artist Drive, and along the range front south of Natural Bridge.

Photo 2. Western front of the Black Mountains, view looking north. The Black Mountains fault zone lies at foot of the mountains.

Photo 3. Fault scarps and Badwater Spring. The two steps in the alluvial fan surface are fault scarps. They mark places where slip on the Black Mountains fault zone displaced one part of the fan relative to the other. This movement must have occurred fairly recently because the displaced fan material is itself quite young. See Photo 6B for an aerial view.

Faceted Spurs

Many west-trending ridges, or spurs, in the Black Mountains end abruptly at an elevation of about 450 m (1500 feet). Below and west of the spurs lies the relatively smooth, steep face of the mountain front that is triangular in shape between the bounding canyons. The spur thus appears "faceted," that is, abruptly truncated by the fault zone at the front of the range. Faceted spurs indicate that this lower part of the range has been exposed so recently, by the dominantly normal movement along the fault zone, that the spurs remain essentially uneroded.

Faceted spurs. An abrupt termination of a ridge at the front of a mountain range that was caused by displacement along a fault.

Wineglass Canyons

On the west side of the Black Mountains escarpment, steep-walled incisions extend eastward into the mountain front. Many of these incisions qualify as wineglass canyons (Photo 4). When viewed from the valley floor, at right angles to the escarpment, the canyons resemble the cross section of

Wineglass Canyon. A canyon whose shape resembles a wineglass, with a narrow, steep mouth and wider, flaring walls towards the back. The alluvial fan at the front resembles the base; the mouth resembles the stem; the wider up-canyon region resembles the bowl.

Photo 4. Aerial view of Black Mountains front, including the Artist Drive area, view looking eastward. The low-lying area covered by alluvium is a small graben, as it is bound on both sides by inward-dipping normal faults. The three canyons in background are wineglass canyons to suggest recent fault activity.

a wine glass. An alluvial fan forms the base of the wine glass, the narrow slot above is the stem, and the higher, wider part is the bowl of the glass. The bowls commonly display a steep lower part and a flaring, or less steep, upper part. Many of the wineglass canyons have a dryfall at the mouth, the bottom of the near-vertical stem.

The stem of the wine glass records the most recent, and continuing interval of down-dip movement on the frontal fault. The vertical walls are the result of rapid down-cutting by stream action. The bottom of the bowl marks the former position of the valley floor and a stable period long enough for the canyon to widen by stream erosion and mass wasting. The less steep, higher part of the bowl is attributable to a still earlier and apparently longer interval of little or no down-cutting, so that the canyon could widen to a greater degree than during the preceding interval.

Particularly good examples of wineglass canyons along the Black Mountains escarpment are at Gower Gulch, behind Artist Drive, and at Tank Canyon, 0.5 km (0.3 mile) south of Natural Bridge. Numerous others can be seen between Badwater and Mormon Point. Titus Canyon along the front of the Grapevine Mountains is still another.

Smaller Faults

Much of the bedrock along the Black Mountains front is broken or crushed by movement along the frontal fault zone. Small-scale faults parallel the frontal fault zone and probably formed during movement on it. Where well exposed, these faults display striated surfaces parallel to the direction of movement, formed by abrasion during slip. The striations on many of these faults trend obliquely and suggest strong lateral motion during the uplift of the Black Mountains.

Turtleback Surfaces

At three places along the west face of the Black Mountains the mountain front lacks the nearly planar appearance produced by faceted spurs, and displays the well known and much debated "turtleback" surfaces. One such location lies immediately north of Badwater, another immediately south of Copper Canyon, and the third at Mormon Point (fig. 3B). Each of these surfaces (1) is convex upward and so shaped like a turtle's shell; (2) curves northeastward into the mountains; and (3) is underlain by a core of the ancient basement complex mantled by younger metamorphosed units consisting mostly of marble (Photo 5A, B). Each is marked by the letter "t" on figure 3B. The mantling rocks have been correlated with the Noonday Dolomite, although at Badwater, Miller (2000) argued that they consist of the lower part of the Crystal Spring Formation of the Pahrump Group. Both the complex and the overlying younger metamorphic rocks have been broadly folded about northwest plunging axes.

The turtleback surfaces owe their identity to the fact that they are also fault surfaces exposed so recently that they remain essentially uneroded. The faults themselves are beautifully exposed along the base of each turtleback surface where they place the cores of the ancient metamorphic rocks against overlying Cenozoic igneous and sedimentary rocks. Like the other faults along the Black Mountains front, the turtleback faults are **normal faults** (Wright et al., 1974). The turtleback faults differ from the range front faults, however, in that they form surfaces against which many faults in the overlying rocks terminate. In this way, the turtleback faults allow the overlying rocks to deform independently of the underlying rocks and so are true **detachment faults**. The distinctive turtle-shell shapes reflect large folds in the metamorphic rocks below the faults, enhanced by the tendency of the turtleback faults to parallel the foliation in the folded metamorphic rocks (Miller, 1991).

The metamorphic rocks beneath the turtleback faults show an array of features to indicate that they have been severely deformed at high temperatures, far beneath the earth's surface. Most prominently, they contain abun-

Normal fault. A type of dip slip fault in which the block of rock beneath the fault surface (the footwall) has moved upwards relative to the block above the fault surface (hanging wall). High-angle normal faults are those in which the fault surface is inclined at an angle of greater than 45°. Low-angle normal faults are those in which the fault surface is inclined at an angle of less than 30°.

Detachment fault. A type of low-angle normal fault in which deformation of the block above the fault surface occurred independent of the deformation below the fault.

Photo 5A. Aerial view of the Badwater turtleback, view looking northward. The high ridge in the background, as well as the low hills at the mountain front, are Late Cenozoic volcanic rock and sedimentary deposits that lie in the hangingwall of the turtleback fault. The brown- and green-colored rock in the foreground and middleground are ductiley deformed Precambrian metamorphic rocks that lie in the footwall of the fault. Just left of the photograph's center, a large wineglass canyon cuts deeply into the footwall. Arrow points to canyon described at mile 55.7 of Black Mountains road guide.

Mylonites. Fine-grained metamorphic rocks that display a strong foliation and lineation. The fine grain size results from extreme deformation under hot conditions.

dant rocks called **mylonites**. Mylonites form in settings similar to fault zones, but at high enough temperatures that the rock flows instead of fractures. Many mylonites therefore form as the deep roots of fault zones exposed at the earth's surface. The turtlebacks provide outstanding examples of this phenomenon, where the metamorphic rocks record their own history of uplift, from relatively deep crustal levels to the surface, along downward projections of the turtleback faults (Miller, 1999a). Estimates for the initial depth of the turtleback mylonites exceed 15 km (Holm et al., 1992; Whitney et al., 1993).

Most recently, Miller and Pavlis (2004) noted that the metamorphic core of each turtleback plunges southeastward as well as northwestward beneath non-metamorphic rocks (Photo 5B). They also noted that the ductile shear zones of each core were probably active at the same time, prior to intrusion of the Black Mountains plutons, at about 10–11 Ma. They proposed that the

Photo 5B. Aerial view of Mormon Point Turtleback (MP) and Copper Canyon Turtleback (CC) looking north. Here, these turtlebacks show the fold-like geometry that resembles the backs of turtles (Curry, 1938). In the background, the Copper Canyon detachment is clearly visible as the boundary between the greenish metamorphic rocks below and the reddish sedimentary rocks (Copper Canyon Formation) above. The Copper Canyon turtleback in this photo also displays the doubly plunging nature of each turtleback; the dashed white line shows the approximate trace of foliation.

three turtlebacks were part of the same mid-crustal shear zone that was active before 10 Ma (Figure 4A). Continued extension, after intrusion of the plutons, formed the three distinct turtleback fault systems. Slip along these faults caused uplift, cooling, and eventual exposure of the turtleback footwalls (Figure 4 B, C).

Alluvial Fans

As the mountains of Death Valley erode, the debris typically finds its way down cliffs or slopes until it reaches a canyon bottom. From there, it continues down canyon, carried by flash floods or debris flows, until it reaches the canyon mouth. There, as the flow emerges onto the flat, open valley floor and loses its energy, the coarsest debris accumulates. Farther

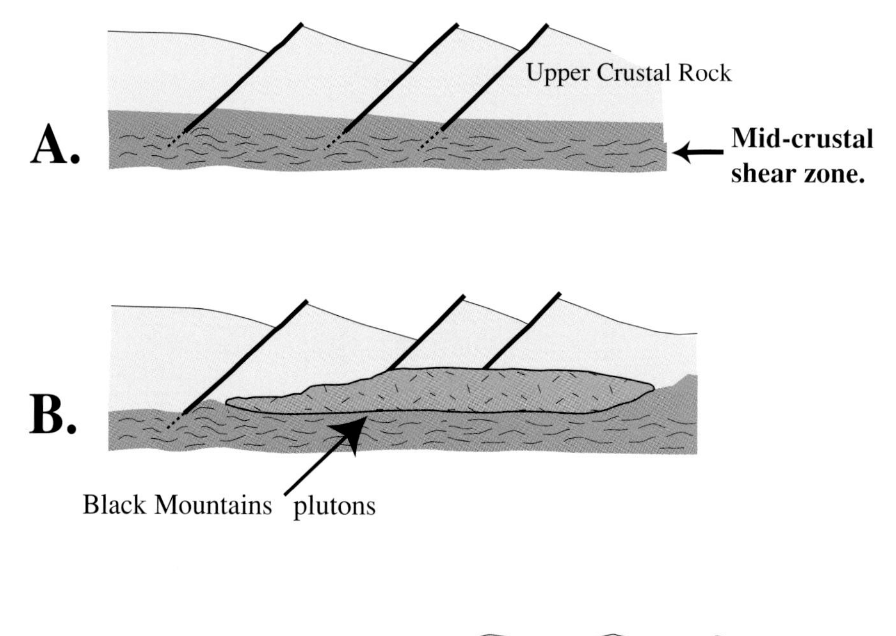

Figure 4. Schematic cross-sections to illustrate principal extensional events at the turtlebacks. 4A: before ~11 Ma, upper crustal faults ended downwards at a mid-crustal shear zone. B: Intrusion of the Black Mountains intrusive suite (Willow Spring Pluton and Smith Mountain Granite) at ~11 Ma. These sill-like plutons intruded along the top of the mid-crustal shear zone. C. Continued extensional faulting and ductile shear exhumed both the intrusive suite and underlying mid-crustal shear zone. Dark-shaded areas represent approximate present-day exposure of metamorphic portion of each turtleback footwall. Cross-sections from Miller and Pavlis (2004).

down the fan, as the flow continues to wane, progressively finer grained material is deposited.

These sedimentary accumulations, beginning at canyon mouths and extending onto the valley floor, are called alluvial fans (Photos 6A, B). Their name stems from the outward radiating stream channels that give a fan-shaped appearance to the deposit. Floods or debris flows frequently encounter clogged channels from previous events and are forced to flow elsewhere. As the new paths eventually become clogged themselves, a series of channels results, each of which starts near the canyon mouth.

In Death Valley, alluvial fans lie along the base of every mountain range, but depending on their specific location, they vary considerably in shape and size. The most striking differences are those between fans on the west side of the valley floor, along the base of the Panamint Range, and those on the east side of the valley floor, along the base of the Black Mountains. Alluvial

Photo 6A. Bajada on west side of central Death Valley. The alluvial fans on the west side of central Death Valley coalesce to form bajadas. Hanaupah fan, shown here, spills out of Hanaupah canyon at an elevation of nearly 900 m (3000 feet) to flow nearly 10 km (6.2 miles) to the valley floor. Note the line of spring-fed vegetation at the bottom of the fan and the prominent fault scarp in the middleground. Telescope Peak is in the background.

fans on the west side coalesce into a broad alluvial slope called a **bajada**. They typically rise well over 300 m (1000 feet) above the valley floor and stretch 5–8 km (3 to 5 miles) in length (Photo 6A). The deposits in some of these fans attain thicknesses as much as 1800 m (6000 feet) (Hunt and Mabey, 1966). By contrast, alluvial fans on the east side and south of Badwater are much smaller: the largest one, below Copper Canyon, measures only 1.6 km (1 mile) in length and less than 400 feet in height. Because they are small, they do not coalesce into bajadas, and so display the fan-shaped geometry very clearly (Photo 6B).

Bajada. *Coalesced alluvial fans along the front of a mountain range.*

Some of this size difference can be explained by the observation that the Panamint Mountains are about twice as high as the Black Mountains and so contribute much more sediment to the valley. However, because the floor of the valley tilts eastward, material on the west side can spread out towards the valley floor, while on the east side, it tends to get trapped at the mountain front (Hunt, 1975).

Photo 6B. Well-shaped alluvial fan at Badwater. In marked contrast with those on the west side, the alluvial fans on the east side of central Death Valley are quite small and well-shaped (also see photo 2). The alluvial fan at Badwater, shown here, attains a height of only 60 m (200 feet) and is less than 900 m (3000 feet) long. The Black Mountains frontal fault lies approximately at the edge of the bedrock. Note the fault scarps, marked by arrows, on both sides of the canyon. Badwater Spring (Photo 3) appears as the green area at the mountain front immediately left of the fan. Liquifaction features, caused by ground shaking during earthquake events, are visible on either side of the road near the outer reaches of the fan, marked by the letter "L" (Wills, 2001).

When viewing an alluvial fan as a whole, it is usually easy to discern which parts are older and which parts are younger. Through time, all exposed rock surfaces in the desert acquire a coating of dark-colored "desert varnish," which consists of manganese oxide. Consequently, the greater the age of a surface, the darker its color (Photo 6A). Moreover, the older fan surfaces typically display larger areas of desert pavement than younger surfaces. Desert pavement appears as a smooth surface of interlocking stones that literally forms a pavement (albeit a fragile one) over the fan (Photo 7).

By comparing slope angles of different parts of the western fans, we see further evidence of ongoing eastward tilting of the valley floor. In general, the older, darker surfaces slope more steeply eastward into the valley (Hunt and Mabey, 1966). Based partly on these relationships, Hooke (1972) suggested that the alluvial fans record six tilting events, three of which post-dated Lake Manly (discussed on page 20).

Photo 7. Desert pavement on bajada near south end of Greenwater Valley. Because it takes a great length of time for desert pavement to develop, it is most prevalent on older, inactive surfaces rather than younger, more active ones.

Badlands Topography

Death Valley National Park contains many outstanding examples of badlands topography, areas eroded into numerous steep-walled gullies separated by narrow ridges. This type of erosion generally affects non-vegetated areas underlain by poorly consolidated, fine-grained materials such as silt, clay, or altered volcanic material. Badlands erosion is enhanced when precipitation occurs as sudden cloudbursts, as opposed to steady, light rains.

The Zabriskie Point and adjacent Twenty Mule Team Canyon areas contain some of the best-known badlands topography in Death Valley (Photo 8). There, countless gullies cut into fine-grained, weakly cemented lake-bed deposits of the Furnace Creek Formation. Alteration of parts of the bedrock to clay allows the surface material to slowly flow downhill when it is wet, thereby enhancing the rate of erosion. Other smaller, but also dramatic examples of badlands exist in the unconsolidated deposits of cinder and ash at Ubehebe Crater, and in the highly altered volcanic rocks in the

Photo 8. Badlands topography at Zabriskie Point. These deeply eroded hills are underlain by lake deposits of the Furnace Creek Formation.

hangingwall of the Badwater Turtleback fault, immediately east of Natural Bridge Canyon.

The Black Mountains and the Basin Ranges from Dante's View

From Dante's View, at the crest of Black Mountains and nearly 1,800 meters (5800 feet) above the valley floor, the visitor gains a perspective of Death Valley that is equally instructive but different from the one obtained from below. To the east and west, the landscape consists of alternating, nearly parallel mountain ranges and intervening basins (Photo 9). This basin-and-range topography characterizes much of the western United States, including all of Nevada, and parts of southern Idaho and southeastern Oregon, western Utah, eastern California, and southern Arizona. Like Death Valley, this topography expresses crustal extension, where mountains rose relative to basins along large normal faults.

Photo 9. View westward from Dantes View. Compare the bajada on the west side of the valley to the small alluvial fan on the east (lower left of photo).

The perspective from Dante's View also shows the asymmetry of the Black Mountains. Their west flank is steep! It drops more than a mile vertically to the floor of Death Valley, which is only about three horizontal kilometers (two horizontal miles) away. By contrast, the east flank descends gradually into Greenwater Valley. This asymmetry reflects the eastward tilting of the range as it rises along the Black Mountains fault zone. On the drive down from Dante's View, for example, one can see to the northeast tilted volcanic flows along the crest of the Black Mountains.

A careful look across the valley at the Proterozoic and Paleozoic sedimentary rocks of the Panamint Mountains reveals that they also tilt eastward. This tilt resulted from normal faulting along the western margin of the range (Fig. 2). From here, one can also easily see how the alluvial fans on the west side of the valley are far larger than those on the east side (Photo 9).

Playa Lakes and the Death Valley Salt Pan

Ephemeral Lakes. Lakes that are filled with water for only short periods of time.

Playa. An exceedingly flat mud-covered surface that occasionally becomes filled with water.

Salt pan. An exceedingly flat salt-covered surface.

Death Valley National Park contains a number of **ephemeral lakes**, called **playas**. Of these the best known are the salt pan of central Death Valley and the Racetrack playa, although large playas also occupy much of northern Death Valley, northern Panamint Valley, and Saline Valley. During rainy periods these exceedingly flat features commonly become "flooded" with less than 2.5 cm (an inch) to more than 30 cm (a foot) of water (Photo 10A). The aridity, however, causes the water to quickly evaporate and leave behind the sediment and dissolved minerals it carried into the playa. Most playas, therefore, are covered by dry, cracked mud commonly associated with evaporites.

The Death Valley **salt pan** is one of the largest modern salt pans on earth. Although its exact boundaries are poorly defined, it extends from the vicinity of the Ashford mill site northward to the Salt Creek Hills, a distance of about 60 km (40 miles). The salt pan is essentially a gigantic, flat sink without a drain. The Amargosa River, which is usually dry, empties into it from the south. The salt pan is also fed from the north by Salt Creek and

Photo 10A. View of temporarily flooded salt pan from Badwater turtleback. Note the reflection of the Panamint Mountains in the shallow water. Brown-colored rock in the foreground is ductiley deformed marble of the turtleback's footwall.

Photo 10B. Salt pinnacles at Devil's Golf Course. Elongated crests of the pinnacles typically trend north-south, parallel to the prevailing wind direction. The vertical column of dust in the background is a dust devil.

from other directions by run-off and spring water. Because this water contains dissolved salts that precipitate as the water evaporates, new salt is continually added to the pan. Much of the salt pan is actually the broad and flat distributary terminus of the Amargosa River. In a strict sense, this part is not a playa, but a low-relief river delta system with alternating channels and flood plain areas.

Visitors to Death Valley can walk onto the salt pan from nearly anywhere along the Black Mountains front. At the Devil's Golf Course one can see pinnacles of salt rising above the surface of the pan (Photos 10A, B). Polygon-shaped blocks, from 1 to 2 meters (3 to 6 feet) across, are produced by desiccation of the salt pan. As the water evaporates and the mud beneath the surface dries, cracks develop between the blocks and are filled with veins of new salt. These cracks ordinarily stand with slight topographic relief above the salt within the polygons. See Photo 10C.

Photo 10C. Polygonal fractures in salt, view towards south. As salt growth tends to concentrate along the broken edges of the polygons, the edges stand out in topographic relief. The Black Mountains front occupies the left background.

Zonation of the Salt Pan

Most of the salts in the salt pan are chlorides, of which halite, ordinary table salt, is the most common. Deposits of sulfate and carbonate evaporites, being less soluble than the salt, precipitate first and so are distributed along the edges of the pan. Hunt and Mabey (1966) showed that the carbonate and sulfate zones are much wider and better developed on the west side of the salt pan than on the east side (Photo 9). They reasoned that the difference was caused by a gentle eastward tilting of the salt pan along the normal faults of the Black Mountains front, allowed the low-solubility minerals to precipitate over a broader, shallow area to the west.

Underneath the Salt Pan

Before Death Valley became a National Monument in 1933, the Pacific Coast Borax Company drilled several exploratory holes into the salt pan in search of potash (Hunt and Mabey, 1966). These holes penetrated the valley fill to depths of as much as 300 m (1000 feet). Individual drill cores showed that the valley fill ranges in composition from mostly salt, through salt plus other evaporites and clay, to mostly clay with minor proportions of evaporites. The salt-rich parts of the cores thus bear evidence of an arid climate much like that of today, whereas the clay-rich parts indicate less arid times when the basin was occupied by a perennial saline lake from which evaporites were occasionally precipitated. Work by Lowenstein et al., (1999) on a more recently drilled core, found evidence for two major arid and intervening wetter periods during the last 200 ka.

Gravity studies tell us about the depth of the salt pan and the configuration of the bedrock below. Strong negative gravity anomalies, for example, indicate the presence of materials of low specific gravity, such as **alluvium** or salt deposits. Original studies of the anomalies show that the valley fill is

Alluvium. Water-transported material.

several thousand feet thick and probably includes Late Tertiary as well as Quaternary strata (Hunt and Mabey, 1966).

More recent gravity studies by Blakely et al. (1999) give a detailed picture of the bottom of the valley fill that supports and extends the earlier research. Their work shows that the seemingly flat floor of Death Valley hides four deeper "sub-basins." From south to north, these sub-basins lie beneath the surface near Mormon Point, Badwater, Cottonball Basin, and Mesquite Flat (Fig. 5). Their maximum depths range from about 3.0 to 3.5 km (1.8 to 2 mi) near Mormon Point to as much as 7 km (4.3 mi) beneath Mesquite Flat. Intervening areas shallow to well less than 1 km (0.6 mi). In general, each sub-basin has steep-walled, probably fault-bounded sides, and probably formed as a result of combined **strike-slip** and **dip-slip** motions on those faults.

Strike-slip faults. Faults in which movement occurred parallel to the strike, or horizontal direction, of the fault plane.

Dip-slip faults. Faults in which movement occurred parallel to the dip, or direction of inclination, of the fault plane.

Racetrack Playa

Racetrack Playa sits at an elevation of 1130 m (3700 feet) in a narrow valley between the southern Cottonwood Mountains and the Last Chance Range (Fig. 3B, Photo 11A). Its surface, which measures more than 4 km (2.5 mi) long by 1 km (0.6 mi) wide, marks a former lake bed that still occasionally floods with shallow water. For the most part, however, Racetrack Playa consists of dry fine sand, silt and clay, broken by countless polygonal mudcracks.

Racetrack Playa is also home to the "sliding rocks." Although no one has witnessed a sliding event, more than a hundred rocks on the playa, which are mostly cobble sized but reach up to 320 kg (700 lbs), rest at the ends of long furrows carved into the hard surface. The average track length exceeds 200 m (650 feet) and displays an irregular, to highly irregular path (Photo 11B; Messina and Stoffer, 2001; 2000). The rocks are especially abundant near the base of a prominent bedrock rib that abuts the south edge of the playa.

Most researchers agree that sliding takes place during high winds when the playa is wet and slippery (see reviews by Sharp and Glazner, 1997; Messina and Stoffer, 2001). Some researchers also call on thin sheets of ice to freeze around the rocks and partially float them as a group on the wet surface, thereby requiring lower wind strengths to cause movement (Stanley, 1955; Reid et al., 1995). Other researchers argue that the expected wind strengths at the playa should be sufficient to cause sliding without ice (Bacon et al., 1996) and that many irregularities in the tracks suggest independent movement (Sharp and Carey, 1976; Messina and Stoffer, 2000). Sharp and Glazner, 1997, eloquently summarize how ice may be an important factor some of the time, but not in all cases.

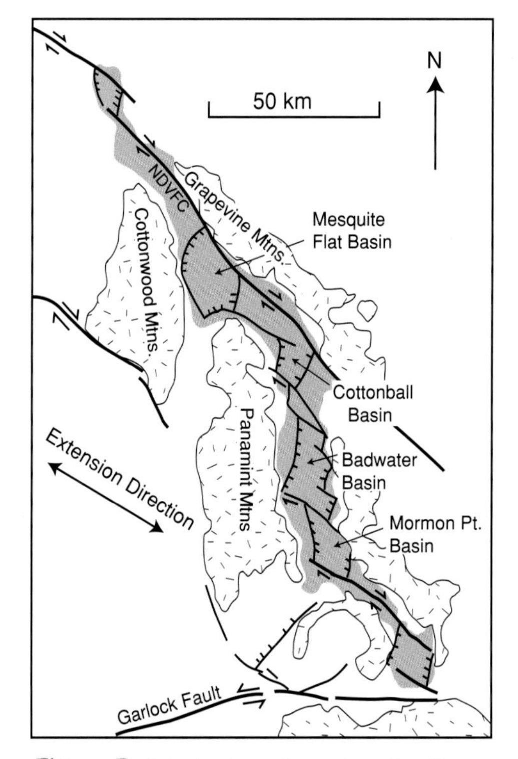

Figure 5. Sub-basins of modern Death Valley as determined by Blakely et al., 1999.

Basics About...

Mining in the Death Valley Area

In 1849, the first people of European descent wandered into Death Valley, having lost their way to the newly discovered gold fields. For the next 80 years, mining dominated the thoughts of most visitors to the region. Even after Death Valley became a National Monument in 1933, it remained open to prospecting and mining for another 40 years. Early prospectors searched out metal deposits, such as gold and silver, but beginning in the 1880s, many prospectors searched out borate and talc. Lingenfelter (1986) provides an unparalleled account of the region's mining and human history.

Photo 3-1. Workings at the Keane Wonder Mine, northern Funeral Mountains. The Keane Wonder Mine produced more than $1 million in gold, mostly between 1907 and 1912. Its mill, however, never reached full capacity because of lack of water.

Metal Mining. Prospectors literally scoured the entire Death Valley region in their search for metals. They found deposits and staked claims within every mountain range, primarily for gold, silver, lead, copper, antimony, and zinc. Few of the mines were successful, hindered largely by the lack of water, but some produced a great deal of ore. The most notable gold districts were Rhyolite/Bullfrog, near present-day Beatty, Nevada, the Keane Wonder mine in the northern Funeral Mountains, and Skidoo, in the northern Panamint Mountains (Photos 3-1, 3-2). Interestingly, the town of Rhyolite, which failed because it ran short of profitable ore in about 1910,

Photo 3-2. Rhyolite, Nevada. Rhyolite's "boom and bust" was as dramatic as probably any mining town in the American west. After gold was discovered in 1904 by Shorty Harris, its population swelled to more than 5000 by 1907 and dwindled to less than 700 in 1910. The schoolhouse, shown in the left background, opened in 1909 when the town was in decline.

lies next to a modern open-pit gold mine. Although now closed, the mine could operate because new technology allowed profitable extraction of gold from much lower grade deposits than before.

Borate and Talc Mining. In contrast to metals, both borate and talc mining proved extremely profitable. During the 1880s, borax was recovered from evaporite deposits scraped from the salt pan of the valley floor. In the early 1900s, large deposits of borate minerals were discovered in Furnace Creek Wash. By the late 1920s, the Death Valley region had become the world's leading source of borax.

The borate deposits in Furnace Creek Wash consist primarily of the minerals colemanite and ulexite-probertite. They reside in lake deposits of the lower 170 m (500 feet) of the Miocene-Pliocene Furnace Creek Formation (McAllister, 1970; Evans et al., 1976). Both open-pit and underground mining were employed to recover the minerals (Photo 3-3). Today, some underground borate mining continues at the Billie Mine, the headframe of which is easily seen on the drive towards Dantes View.

Photo 3-3. Open-pit borate mine, the "Boraxo Pit," Upper Furnace Creek Wash. This part of Death Valley, traversed by tourists on the way to Dantes View, contains some of the world's largest deposits of colemanite and probertite, two important borate minerals. The deposits exist within the Furnace Creek Formation. The region was mined extensively beginning in the 1920's, first in large underground mines, and then in open pits such as this one. Photo 28 shows the borate mining camp of Ryan.

Death Valley's talc deposits reside within dolomite of the Crystal Spring Formation. They formed by contact metamorphism of the dolomite when it was intruded by diabase sills at about 1 Ga (see Photo 23; Wright, 1968). Most talc mining occurred in Warm Springs Canyon of the southern Panamint Mountains and in the Ibex Hills of southern Death Valley. Death Valley talc, which was of especially high purity, commanded approximately 12% of the US market in 1974 (Evans et al., 1975). It was used principally for ceramics, paint extenders, and insecticides. No talc mining presently takes place inside the national park.

Photo 11A. Intrusive contact at Racetrack Playa, view looking northwestward. Here, Jurassic granitic rocks, which make up the Ubehebe Peaks to the left of the arrow, intrude folded Paleozoic sedimentary rocks to the right of the arrow. Racetrack Playa occupies the middle ground; Saline Valley occupies the background.

The bedrock setting of Racetrack Playa is as fascinating as the sliding rocks (Map 1, page 106). On either side of the playa, part of the Jurassic Hunter Mountain Batholith intrudes strongly folded Paleozoic limestone and dolomite, some of which makes up the bedrock rib at the south end of the playa (Photos 11A, B). The batholith, which consists largely of quartz monzonite (a type of granitic rock), underlies much of the southern Cottonwood Mountains, including the nearby Ubehebe Peaks (Fig. 3A). It also

makes up "The Grandstand," a twenty-meter-high outcrop that sticks out of the playa near its northwest corner. The Grandstand rocks are unusual in that they contain many oversized feldspar crystals. At the contact zone between the batholith and Paleozoic rocks, chemical reactions formed a variety of unusual minerals. A good exposure of this zone lies near the top of the trail up Ubehebe Peak.

Lake Manly

At many places in Death Valley we see evidence that the salt pan was once occupied by a relatively deep and extensive body of saline water known as Lake Manly. In fact, according to drill core analysis by Lowenstein et al., 1999, the lake filled Death Valley for two extended periods of time in the last 200 ka. The largest lake existed from 186 to 120 ka and a smaller one existed from 35 to 10 ka. These periods generally corresponded with episodes of glaciation at higher latitudes and elevations.

The most obvious evidence for this lake exists in ancient shorelines, stranded above the valley floor. These shorelines appear as narrow, horizontal benches that are typically covered by beach gravels, and formed when the lake level was stationary long enough for waves to cut them. They are most numerous and best preserved at Shoreline Butte (Photo 12) west of the Ashford Mill site in southern Death Valley. They are also well preserved at Badwater and in the faulted alluvial fans exposed on the northeast side of Mormon Point.

Photo 11B. Sliding rock and track near the south edge of Racetrack Playa.

The elevation difference between the lake's lowest point at Badwater and its highest bench at Shoreline Butte approximates the maximum depth of Lake Manly to be about 180 m (600 feet). However, this difference does not provide an entirely accurate measure of the depth because of eastward tilting along the Black Mountains fault zone since the lake retreated. In fact, Hunt and Mabey (1966) reported that the shorelines on the western side of Death Valley are higher than on the east side because of this tilting.

Other evidence for the lake includes preserved gravel bars and accumulations of calcium carbonate called "tufa," deposited along its shores. An excellently preserved gravel bar is cut by the highway to Beatty, about 2 miles north of Beatty Junction. Nearly horizontal accumulations of tufa mixed with fragments of the underlying bedrock cling to the Black Mountains front, and are especially noticeable near Badwater.

Photo 12. Shoreline Butte. Lake Manly's shorelines are preserved as the many narrow, horizontal benches that reach most of the way to the top of this feature. A spring bloom of yellow wildflowers flanks the butte.

Desert Pupfish

Several species of desert pupfish (Cyprinodon) manage to survive year after year in springs and creeks of Death Valley. These fish, who reach lengths of only about 4 cm (1.5 inches), can withstand temperature variations of 21°C (70°F), as well as wide ranges in water salinity (Brown, 1971). They are relevant to the geology of Death Valley because their evolution tracks the drying of the region's climate and disappearance of Lake Manly.

During the Late Pleistocene, Lake Manly was the largest of several lakes in the region which, together with free-flowing streams between them, formed an integrated drainage network (Fig. 6A; Miller, 1950). Ancestral pupfish populated the system. As the climate dried, however, individual lakes and springs became separated from the network, and with them, groups of pupfish. Living in isolation, different populations of pupfish evolved into different species. Those that became isolated earlier show more striking differences to the other species than those that became isolated later (Soltz and Naiman, 1978).

Today, visitors can see four different species of pupfish at four different localities (Fig. 6B). One of the most accessible is Salt Creek, on the valley floor just north of Furnace Creek. This

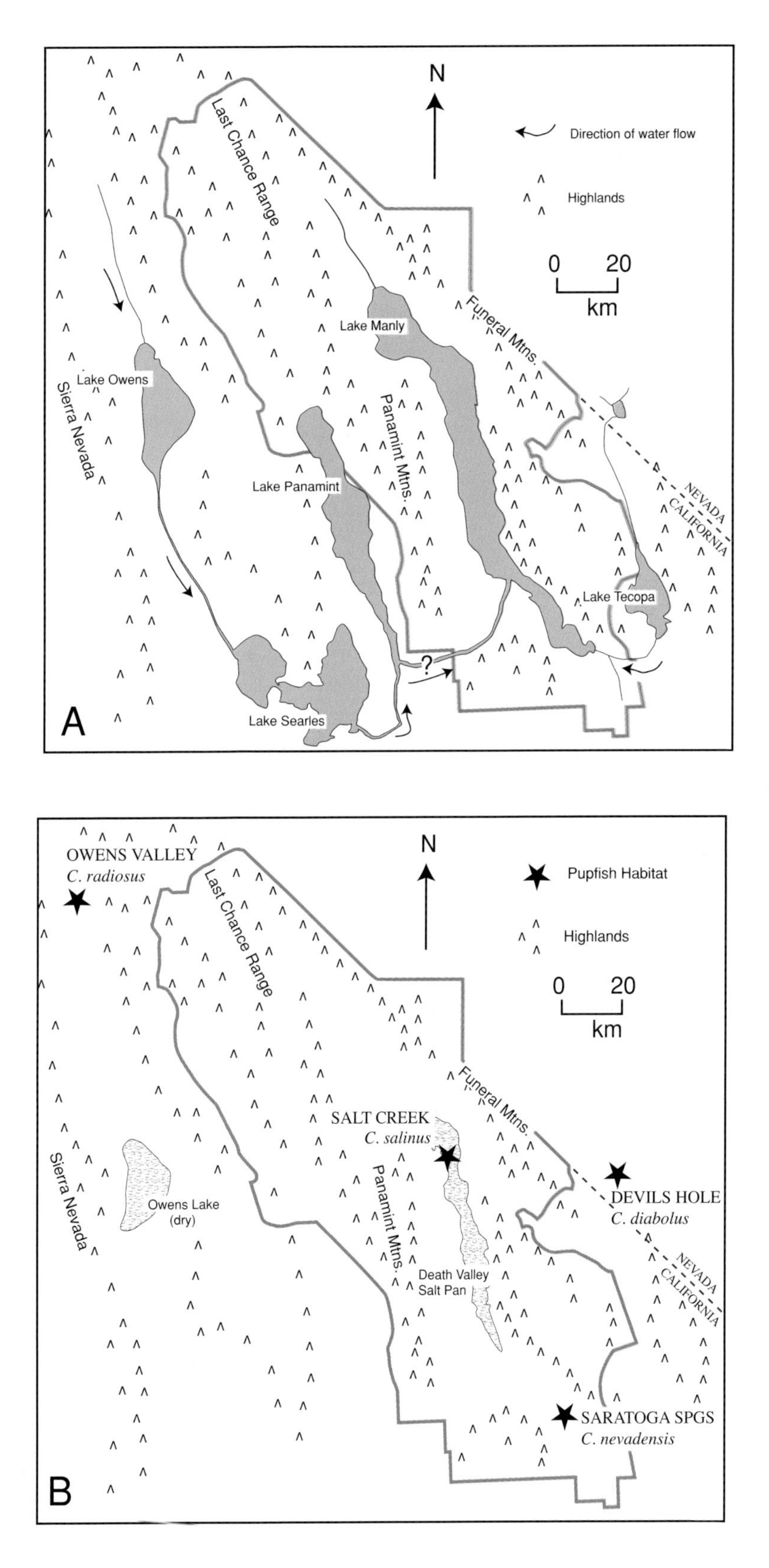

Figure 6. Surface water in the Death Valley region during the Late Pleistocene and today. 6A, Integrated drainage system in the Death Valley region during the late Pleistocene. Question mark near Wingate Pass highlights uncertainty in this locality as a direct connection between Lakes Panamint and Manly (Machette et al., 2001b). 6B, Today's lack of surface water and isolated pupfish localities. Data compiled from Brown, 1971; Snyder et al., 1964; Miller, 1950.

locality, which hosts the species *Cyprinodon salinus,* is also one of the harshest, as the creek dries in mid-summer to a series of marshes. The Amargosa Pupfish (*C. nevadensis*) inhabits springs and pools along the Amargosa River; a subspecies of this fish occupies Saratoga Springs in the south end of the park. Devils Hole Spring hosts *C. diabolis*, the smallest and longest isolated of the desert pupfish. Because the spring at Devils Hole lies deep within faulted limestone it receives little sunlight each day. The resulting limited supply of algae restricts the population of this pupfish to only 200–300 individuals. In the Owens Valley, a population of *C. radiosus* survives in protected habitats near Bishop.

Young Volcanic Features

Evidence of recent volcanic eruptions in Death Valley can be viewed at two easily accessible localities within the park. In southern Death Valley, near its confluence with Wingate Wash, a basaltic cinder cone less than 400,000 years old projects above the valley floor. At the northeast edge of the Cottonwood Mountains, Ubehebe Crater (Photo 13A) and several smaller craters formed as recently as 300 years ago (Klinger, 2001a).

Photo 13A. Aerial view of Ubehebe and adjacent craters, view looking north-eastward. The eruption at Ubehebe crater was the largest and most recent of at least 13 crater-forming eruptions (Crowe and Fisher, 1973). This photograph shows seven of those craters. A smaller chain of craters (not seen in this photograph) extends from the main crater westward for about 600 m (2000 feet).

The cinder cone in southern Death Valley appears as two red hills, easily visible from the highway southeast of Mormon Point. The hills mark the opposite sides of a cone, originally intact, but now offset several hundred meters by right-lateral movement on a strand of the Southern Death Valley fault zone (Photo 13B). The cinder cone also coincides with a normal fault zone, which, in a seismic reflection line, appears to connect the cinder cone to a body of **magma** at a depth of about 15 km (9.3 mi) (de Voogd et al., 1986).

Magma. Liquid rock that is beneath the earth's surface.

Ubehebe Crater ranges from about 150–200 meters (500 to 700 feet) deep, is half a mile wide, and lies on the edge of numerous other smaller craters. The craters resulted from explosions generated when rising basaltic magma contacted ground water and flashed it to steam. Small explosions that formed the smaller craters were followed by a much larger explosion resulting in Ubehebe Crater. No lava was extruded, but up to 50 meters (150 feet) of black cinders and ash blanket an area of about six square miles. Highly colored alluvial deposits underlie the cinders and are exposed in the walls of Ubehebe Crater. The colors are best attributed to oxidation just prior to the eruptions by percolating hot ground water.

Sand Dunes

Dune fields are widely distributed in and around Death Valley National Park. Contrary to a common perception, however, they occupy a very small fraction of the total area of the park. The most frequented, most photo-

Photo 13B. Aerial view of cinder cone in southern Death Valley, offset right-laterally by a strand of the southern Death Valley fault zone. The West Side Road, which extends diagonally across the photograph, allows easy access.

Photo 14A. Wind ripples on sand dunes at Mesquite Flat, adjacent to Stovepipe Wells.

graphed and also one of the largest occupies part of Mesquite Flat north of Stovepipe Wells (Photo 14A). It is most easily accessed from State Highway 190. Equally photogenic but less accessible dune fields include those near Saratoga Springs in southern Death Valley, in northern Panamint Valley, Saline Valley, and in southern Eureka Valley. Just east of the national park boundary lie the Big Dune and Dumont dune fields.

Sand Dunes. Large accumulations of wind-blown sand.

In order to form, **sand dunes** require a steady supply of sand, wind to blow the sand, and a wind break that allows deposition of sand. Each dune field in Death Valley National Park receives most of its sand from the nearby alluvial fans and experiences plenty of wind. Furthermore, each dune field lies in an area that is slightly sheltered from prevailing winds. The dunes at Stovepipe Wells, for example, lie in the embayed mountain front just north of Tucki Mountain while the dunes in the Eureka Valley lie between the Last Chance and Saline Ranges (Photo 14B). In general, dunes of the Death Valley area tend to form close to their source areas. They are small in the central part of Death Valley as much of the sand there is cemented in the salt pan (Hunt and Mabey, 1966).

Photo 14B. Eureka Sand Dunes, as viewed from the Last Chance Range, towards the south. These dunes, which rise over 183 m (600 feet) above the valley floor, are the highest dunes in California. The dune field is about 5 km (3.1 mi) long.

Dunes can be classified according to their shapes. Star dunes consist of three or more arms and generally require multiple prevailing wind directions to form. Crescentic and linear dunes both contain two arms but differ in that crescentic dunes are asymmetrical in profile while linear dunes are symmetrical. At Mesquite Flat, the dune field contains both star and crescentic dunes (Photo 14C). Low, flat **interdune areas** lie between the crescentic dunes. The star dunes tend to be high and to lie near the center of the field whereas crescentic dunes are lower and lie closer to the margins. At the far margins of the field, stands of mesquite grow atop stable conical hills of sand.

Interdune areas. Patches of bare ground between sand dunes.

Remnants of a sheet-like deposit of fine-grained, mudcracked sediment occupy much of the interdune areas at Mesquite Flat. This deposit accumulated on the bottom of a lake that recently covered the area now occupied by the dunes (Photo 14D). Presently, this deposit is being eroded in some places, and buried by sand in others.

Springs

Parts of Death Valley have a surprising abundance of spring water. Springs in the area of Furnace Creek Wash, for example, discharge approximately 6000 liters (2000 gallons) per minute, easily enough to supply the

Photo 14C. Aerial view of crescentic and star dunes at Mesquite Flat, looking eastward. Most of the dunes are crescentic dunes, but several large star dunes, each with three or more arms, are visible on the north (left) side of the field.

Photo 14D. Exposure of former lake bed at Mesquite Flat. Similar exposures of this former lake bed show up between many of the crescentic dunes.

park village. Elsewhere in Death Valley, particularly along the mountain fronts, springs provide most of the water that supports its diverse and fascinating ecology.

In Death Valley, as elsewhere, springs mark the places where ground water, flowing from higher to lower levels, reaches the surface of the land. Most of the springs in the Death Valley region emanate from fault zones or the toes of alluvial fans (Hunt, 1975). Fault-controlled springs abound along the fronts of the Grapevine, Funeral, and Black Mountains. Some of the more accessible of these are the spring at Badwater (Photo 3), Travertine Spring in Furnace Creek Wash, and the Keane Wonder Spring at the front of the Funeral Mountains just north of the Keane Wonder mine (Photo 15A). Klare Spring, along the road down Titus Canyon in the Grapevine Mountains, issues from a fault zone within the mountains.

Water discharges from the toes of alluvial fans where the highly permeable sand and gravel of the fan grades abruptly into the less permeable sand, silt, and clay of the bordering playa. Because water cannot easily penetrate the fine-grained sediment, it flows out to the surface of the fan. The alluvial fan at the mouth of Furnace Creek is a spectacular example of this kind of

Photo 15A. Spring deposits along Keane Wonder Fault, northern Funeral Range, view looking southward. These deposits consist largely of travertine, a form of calcium carbonate precipitated from the spring water.

Photo 15B. Springs on the Furnace Creek alluvial fan, looking northeastward. Here, numerous springs, which emerge radially on the lower reaches of the fan, are marked by lines of vegetation.

spring environment. There numerous spring-fed lines of vegetation originate at the fan-to-playa transition and radiate outward onto the playa (Photo 15B). Many other springs, such as Shorty's Well and Bennett's Well, lie at the toes of the fans that slope eastward from the Panamint Range.

The springs that supply water to Salt Creek, on the floor of northern Death Valley, formed differently. North of the springs, ground water moves easily through the permeable alluvium on the floor of the valley. The springs mark the places where the ground water rises upon encountering the impermeable fine-grained strata of the Miocene-Pliocene Furnace Creek Formation, the rock that underlies the Salt Creek Hills. Southward and downstream from there, Salt Creek disappears back into the more permeable valley fill material.

Most of the water that feeds fault-controlled springs in the Funeral and Grapevine Mountains comes from central Nevada. It flows through carbonate aquifers at depths that range from 0 to 500 meters (1600 feet) before discharging at the surface. Devil's Hole, a fault-controlled spring on the east edge of the Amargosa Valley, provides a "window" into the carbonate aquifer. There, water discharges from a cave in carbonate bedrock but does not flow out over the surface and escape. Instead, it forms a pool that provides the sole habitat for the species of pupfish *Cyprinodon diabolis*. Most of the water that does not evaporate continues on its journey westward to the springs of Death Valley.

Ventifacts

The floor of Death Valley, as well as floors of most of the adjacent valleys, displays graphic evidence for wind erosion in the form of sand-blasted rocks called ventifacts (Photo 16). These rocks are faceted and typically display parallel grooves carved by blowing sand. Some spectacular ventifacts exist on the low ridge opposite the turn-off to Artist Drive, and north of Highway 190 in Panamint Valley. Interestingly, the unusually shaped rock on the Badwater Road named "Mushroom Rock" is not a ventifact. Instead, it probably attained its shape through the weathering effects of salt (Meek and Dorn, 2000).

Photo 16. Ventifacts on small ridge, just opposite turn-off to Artist Drive. Note also the well-developed desert pavement.

Chapter Two

Shaping of the Present Landscape:

Extensional Tectonics and the "Basin and Range Event"

View northward along the Black Mountain front. Dark green rock in the foreground is the footwall of the Copper Canyon Turtleback.

First-time visitors to Death Valley National Park invariably and correctly sense that these mountains and valleys have formed quite recently in geologic time. Beginning about 16 million years ago, this part of the earth's crust was broken into a gigantic mosaic of mountain blocks each bounded by major faults. Today's youthful appearance of the Death Valley landscape suggests that slip on these faults is still happening. Specific features of the Black Mountains rangefront that indicate its youthfulness are discussed in Chapter 1.

Cenozoic sedimentary and volcanic rocks deposited during the development of the present landscape provide a detailed record of extension in Death Valley. Many of these deposits are preserved within the intervening sedimentary basins, but some are exposed within the ranges (Fig. 3A). Among other things, these deposits tell us that Death Valley itself formed during the latest period of crustal extension. They also indicate that a precurser to Death Valley, called the Furnace Creek Basin, dominated the landscape from about 16 Ma to about 4 Ma.

The Faults

In the faults that bound the ranges and basins of the Death Valley region (Fig. 3A, B), as in the rest of the Basin and Range province, we find visible evidence of an extending crust. The faults are the principal ruptures along which the brittle, upper part of the crust has broken as the great block of the Sierra Nevada moved westward away from the west side of the Colorado Plateau. The land between the two has been literally pulled apart. The study of fault patterns generated in this way belongs to a branch of geology called extensional tectonics.

Classification

The faults that presently define the ranges and valleys of Death Valley National Park (Figs. 3A, B) are broadly divisible into three kinds: strike-slip, high-angle normal and low-angle normal. Most of the ranges also contain older thrust faults that formed during a much earlier period of crustal contraction. The strike-slip faults, along which movement has been dominantly parallel with the strike of the fault planes, are identified by arrows on Figure 3A that show sense of lateral movement. Note that most of these faults strike northwestward and that their southwest sides have moved relatively northwestward, producing a "right-lateral" sense of slip. Movement on the faults identified as normal has been mainly down-dip. The normal faults, by virtue of their geometries, are the simplest expressions of crustal extension; the lower the angle, the greater the opportunity for large-scale extension. Basics About Faults and Fault Zones describes these different fault types in more detail.

Most students of the structural framework of the Death Valley region view the major range-bounding faults as terminating at depth against nearly horizontal "detachment zones." In this way, deformation above the fault occurs independently of deformation below the fault. Some researchers place the principal detachment surface beneath much of the Death Valley region at mid-crustal levels (Serpa et al. 1988); others favor a much shallower depth (Wernicke et al. 1988).

High-angle normal faults. As described in the introduction and first chapter, much of the present-day topography of Death Valley is the result of slip on normal, or oblique-normal faults that dip westward at angles of 45° or more (Fig. 2). As these faults slip, they cause the adjacent rocks to rotate down towards the east. This combination of fault slip and rotation produces mountain ranges with steep, fault-bounded western margins and relatively gentle eastern sides. The west face of the Black Mountains stands out as one of the world's most spectacular examples of this type of fault-generated topography. The western margins of the Panamint and Cottonwood Mountains also provide outstanding examples of fault-bounded mountain fronts.

Low-angle normal faults. Much of the extension in Death Valley has been controlled by slip along low-angle normal faults. These faults are frequently called "detachment faults" because they allow their upper plates to deform independently of their lower plates. Some of the better-known and accessible detachment faults are the Boundary Canyon fault in the northern Funeral Mountains, the Turtleback faults and the Amargosa fault in the Black Mountains, and the Mosaic Canyon fault in the northern Panamint Mountains. The Turtleback faults are described in detail on page 5.

The Boundary Canyon fault is one of the most conspicuous of Death Valley's low-angle normal faults (Photo 17). It dips gently northwestward from the northern Funeral Mountains to beneath the highly folded and faulted formations of the Grapevine Mountains. Thus the Grapevines lie in the upper plate, and all but a small part of the Funerals lie in the lower plate. In the northern Funerals the lower plate consists of Proterozoic rocks, including all three formations of the Pahrump Group, the Johnnie Formation and the lower part of the Stirling Quartzite (Table 1). The upper plate contains the upper part of the Stirling Quartzite and all of the overlying formations through those of Mississippian age (Wright and Troxel, 1993).

The Boundary Canyon fault is of special interest in that the rock units of the lower plate have been metamorphosed at temperatures and pressures that characterize mid crustal levels, whereas the rocks of the upper plate remain essentially unmetamorphosed. The components of such a geologic setting are commonly and collectively called a metamorphic core complex and require large displacement along the low-angle normal fault that separates the two plates. Hoisch and Simpson (1993) estimated that the upper plate moved about 40 km (25 miles) along the fault towards the northwest.

The Boundary Canyon fault is easily discernable along the west face of the Funeral Mountains east of the highway as one approaches the mouth of Boundary Canyon from the south. It is also exposed on both sides of the lower part of the canyon. In this

Photo 17. Boundary Canyon Fault, northern Funeral Mountains. The Boundary Canyon fault separates the greenish rocks below from the tan-colored ones above. The lower rocks, which belong to the Johnnie Formation, were brought up from mid-crustal depths along the fault. As a result, they show features indicative of metamorphism and deformation at high temperatures. The overlying rocks, however, which belong to the Stirling Quartzite, were deformed at much lower temperatures and are only slightly metamorphosed.

area the fault dips gently to the northwest and separates light colored, unmetamorphosed strata of the middle part of the Stirling Quartzite from drab exposures of strongly metamorphosed and deformed units of the middle and lower parts of the Johnnie Formation (Table 1). Equally metamorphosed rock units of the underlying Pahrump Group are superbly exposed in nearby Monarch Canyon, which drains westward from the crest of the Funeral Mountains.

Chaos. A structural term for a mosaic of fault-bounded, typically gigantic blocks, derived from a stratigraphic succession and arranged in proper stratigraphic order, but occupying only a small fraction of the thickness of the original succession. In the Death Valley region, where LF Noble coined the term, chaos in normally viewed as a product of extreme crustal extension.

The Amargosa Fault and the Amargosa Chaos

Of the extension-related phenomena of the Death Valley region, one of the best known, most complex and most controversial is the Amargosa chaos, first described by Noble (1941) and mapped in detail by Wright and Troxel (1984). Noble originally recognized three phases of the chaos, but the one that he named the "Virgin Spring phase" is now viewed as true **chaos** in the sense that he introduced the term. It is exposed in separate localities in the southern Black Mountains from near Virgin Spring and Rhodes Washes northward to Gold Valley (Photo 18).

In simplest terms, the chaos consists of a mosaic of fault-bounded blocks of Proterozic and Cambrian formations, arranged in proper stratigraphic order, but highly attenuated to a small fraction of the actual combined thickness of the formations represented (Fig. 7). Most of the chaos rests in fault contact upon intact occurrences of the Early Proterozoic crystalline complex. In some places within the Chaos, however, the base of the Proterozoic sequence rests depositionally on the basement (Wright and Troxel, 1984). All of the faulting occurred in the shallow crust. In the Gold Valley area, in the central part of the Black Mountains, the chaos is intruded by granitic bodies that predate 10 million year old volcanic units. Thus the chaos may be the oldest extension-related structural feature in the Black Mountains block. An excellent exposure of the chaos lies immediately south of the highway at a point 2.4 km (1.5 miles) east of Jubilee Pass.

Noble (1941) originally interpreted the chaos as remnants of a single, region-wide **thrust fault** and named by him the "Amargosa thrust." Because it thins, rather than thickens the stratigraphic

Thrust fault. A type of dip slip fault in which the block of rock beneath the fault surface (the footwall) has moved downwards relative to the block above the fault surface (hanging wall). Thrust faults typically bring older rocks over the top of younger rocks.

Photo 18. The Amargosa Fault and the Amargosa chaos. The Amargosa Fault (heavy dashed line) separates crystalline basement rock (p€b) from overlying sedimentary rock. The overlying sedimentary rock displays such a complex style of faulting that it is called "chaos." Two of these faults appear in the photograph as red dashed lines.

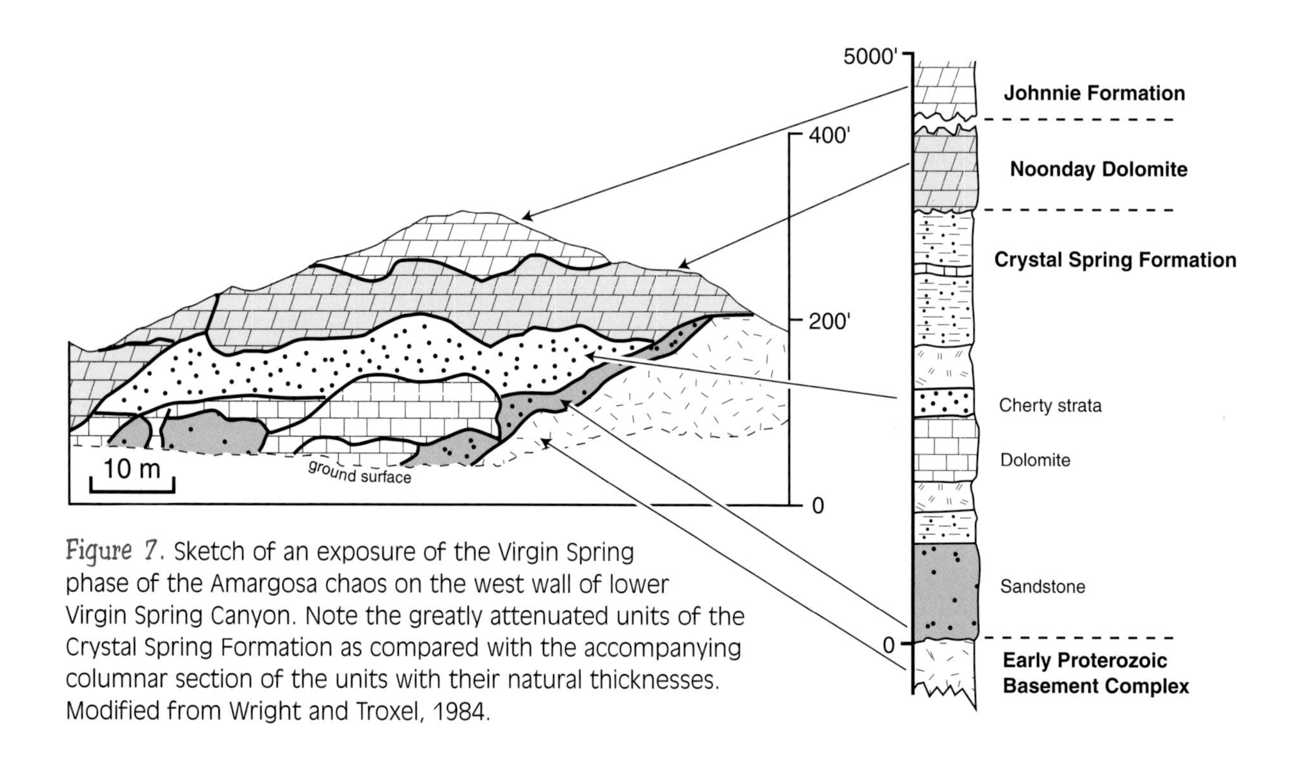

Figure 7. Sketch of an exposure of the Virgin Spring phase of the Amargosa chaos on the west wall of lower Virgin Spring Canyon. Note the greatly attenuated units of the Crystal Spring Formation as compared with the accompanying columnar section of the units with their natural thicknesses. Modified from Wright and Troxel, 1984.

section and formed during the late Tertiary, the Amargosa fault is now viewed as composing segments of one or more low-angle normal faults. The Amargosa Chaos has been so disordered by later faulting and folding as to make its original configuration extremely difficult to accurately reconstruct.

Strike-Slip faults. Strike-slip faults have long been recognized as prominent structures throughout the region, as they are present in nearly every large valley between the Sierra Nevada and Las Vegas, Nevada (Wright, 1989). In Death Valley, strike-slip faults control the modern-day crustal extension, and likely exerted a controlling influence on earlier periods of extension.

The Death Valley Fault System and the Origin of Modern Death Valley

Figure 7 shows the active Death Valley fault system in solid red lines. It consists of three parts: the Northern Death Valley fault zone (NDVFZ), the Black Mountains fault zone (BMF), and the Southern Death Valley fault zone (SDVFZ) (Machette et al., 2001). Each of these faults shows abundant evidence for recent activity, including offset stream channels and fault scarps (Klinger, 2001b, Brogan et al., 1991); details of features along the Black Mountains fault zone are described in an earlier section of this book. This fault geometry, where the northern and southern Death Valley fault zones are linked by the Black Mountains fault zone, drives the modern extension in

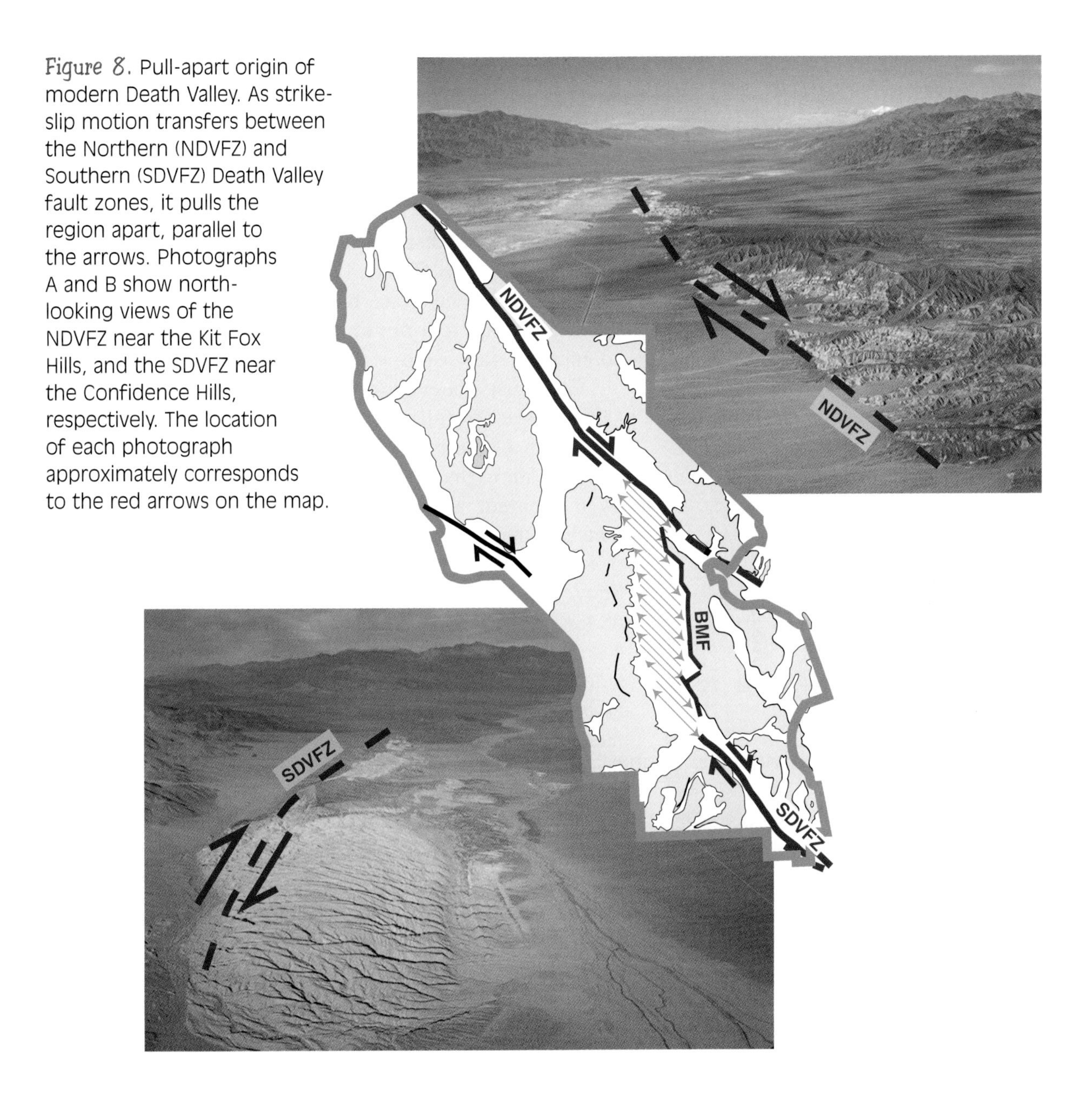

Figure 8. Pull-apart origin of modern Death Valley. As strike-slip motion transfers between the Northern (NDVFZ) and Southern (SDVFZ) Death Valley fault zones, it pulls the region apart, parallel to the arrows. Photographs A and B show north-looking views of the NDVFZ near the Kit Fox Hills, and the SDVFZ near the Confidence Hills, respectively. The location of each photograph approximately corresponds to the red arrows on the map.

Death Valley. Central Death Valley is therefore a "pull-apart" basin, as its crust is being pulled apart between the two strike-slip fault zones (Fig. 8; Burchfiel and Stewart, 1966).

The Furnace Creek Fault Zone

The best known of the ancient strike-slip faults exposed within the park boundaries compose the Furnace Creek fault zone (Figs. 3A, B). This fault zone defines a linear crustal rupture that extends from the vicinity of Eagle Mountain northwestward for about 250 km (150 miles), including Furnace

Photo 19. Aerial view of the Furnace Creek fault zone near Hole-in-the-Wall, Furnace Creek Wash. Here, the fault separates Paleozoic rock (Pz) of the Funeral Mountains from the Tertiary Furnace Creek Formation (Tf).

Creek Wash and the full length of northern Death Valley. North of Furnace Creek, it coincides with the active, northern Death Valley fault zone (Machette et al., 2001a). South of Furnace Creek, it trends southeastward up Furnace Creek Wash as the dashed red line of Figure 8. There, it brings Miocene and Pliocene strata of the Furnace Creek Basin into contact with the Proterozoic and Paleozoic formations of the Funeral Mountains (Photo 19; McAllister, 1970). The Miocene and Pliocene strata, being as much as 15,000 meters thick, also require a minimum vertical displacement of a comparable dimension.

Geologists, beginning with Stewart and his coworkers (1968), have carefully inspected the pre-Cenozoic rocks on both sides of the Furnace Creek fault zone, and have observed features that were once joined and are now separated by movement along the fault zone. In making these matches, they find compelling evidence for displacements of tens of kilometers. Along the southern part of the fault, Snow and Wernicke (1989) and Snow (1992) argued for 68 km (42 mi) of right-lateral slip based on their proposed correlation of pre-Tertiary thrust faults and folds on either side of the fault. Stevens et al. (1991, 1992) suggested 80 km (50 mi) of right slip, based on an offset sedimentary facies boundary in Mississippian-aged rock as well as a different correlation of thrust faults and folds.

Other important strike-slip faults in Death Valley include the active Hunter Mountain fault in the northern Panamint Valley, and the inactive Sheephead fault, in the southern Black Mountains. The Hunter Mountain fault zone offsets the Hunter Mountain batholith right-laterally by 8–10 km (5–6 mi) (Burchfiel et al., 1987). The Sheephead fault bounds the southern edge of magmatic rocks in the Black Mountains and is probably also right-lateral (Wright et al., 1991). The Sheephead fault is enigmatic, however, because it is mostly obscured by alluvium. Additionally, the Southern Death Valley fault zone, which forms the southern edge of the modern pull-apart basin, shows evidence for a history that dates back at least several million years. Along this fault, alluvial fan gravels are offset right-laterally about 35 km (22 miles) from their source in the southern Panamint Mountains (Butler et al., 1988).

The Ranges

Although the mountain ranges within Death Valley National Park have formed in a framework of interrelated Cenozoic faults and generally qualify as **fault-block ranges** (Fig. 2), some differ widely in other respects. Most importantly, they differ in the specific types of bedrock of which they are composed, and in the types of faults that control their internal structure.

Most ranges are dominated by the striped outcrops that identify the evenly bedded Proterozoic and Paleozoic formations (Photo 20). These ranges include the Panamint, Cottonwood, Last Chance, Funeral, and Grapevine Mountains. Of these, the Funeral and Grapevine Mountains stand out because they lie end-to-end to form a single, northwest-trending topographic high, adjacent to the Furnace Creek fault zone. This composite range has extended internally along normal faults oriented approximately perpendicular to its backbone. The transition between the two ranges coincides with the Boundary Canyon low-angle normal fault, in that the Grapevine Mountains lie in the hangingwall and the Funeral Mountains lie in the footwall of the fault (Photo 17).

The Black Mountains and Greenwater Ranges are underlain largely by irregular bodies of Tertiary plutonic and volcanic rocks. They also contain most of the park's exposures of crystalline basement rock, mostly in the southern Black Mountains, but also in the cores of the three turtlebacks. As a consequence, these ranges exhibit a certain variability in their appearance:

Fault-block range. A mountain range that has risen along a master fault zone. Most ranges in Death Valley are tilted fault blocks, because they have risen and tilted along normal faults.

Photo 20. View looking southeastward into Death Valley from Aguereberry Point. In the foreground, Paleozoic rock dips eastward. The prominent white unit is the Cambrian Zabriskie Quartzite.

the Tertiary intrusive and crystalline basement rock present a somewhat dull and somber appearance whereas the Tertiary volcanic rocks tend to be brightly colored.

The Owlshead Mountains, in the southern part of the park, differ from the other ranges in that they are equidimensional in map view, and essentially coextensive with a cluster of granitic **plutons** of Mesozoic age (Fig. 3A). The plutons are discontinuously covered by Miocene **andesitic** and rhyolitic lava flows and are also offset by strike-slip faults. Most of these faults strike northeast and have moved in a left-lateral sense.

Plutons. Bodies of intrusive rock.

Andesitic. A compositional term used to describe magmas or volcanic rocks that consist of between 60-65% silica. Of or like the rock andesite.

The Basins

Cenozoic sedimentary deposits, which have accumulated in the topographic depressions between the ranges of the Death Valley region, consist mainly of debris eroded from the high areas and deposited in alluvial fans, ephemeral and **perennial lakes**, and stream beds. Also included are accumulations of evaporites, principally limestone, gypsum and salt, brought in solution by streams. The sedimentary fill is typically interlayered with extrusive volcanic rocks.

Perennial Lakes. Lakes that are constantly filled with water.

By observing the shapes of these basins and the distribution of the various kinds of sedimentary rocks that they contain, by dating the extrusive volcanic bodies in the basins, and, on occasion, by employing geophysical methods to detect subsurface features, geologists can reconstruct the development of the basins and the erosional history of the source areas. The task is hindered by a cover of Quaternary alluvium that hides much of the older Cenozoic deposits. Late Cenozoic faulting and folding, however, has exposed to erosion the pre-Quaternary rocks of several basins within the park boundaries.

The Cenozoic sedimentary basins of the Death Valley region have evolved in a variety of ways, each in response to the interplay of the three major types of faults discussed above. Deposits of three basins are especially well displayed along the main roads of the national park. In the floor of central Death Valley sediments are accumulating today in the pull-apart basin bounded on the east by the zone of normal faults that defines the front of the Black Mountains (Fig. 3A). In Furnace Creek Wash, a thick succession of Miocene to Pliocene sedimentary deposits, volcanic ash, and lava flows defines the Furnace Creek Basin. Because it lies adjacent to the Furnace Creek fault zone, this basin has been called a "strike-slip" basin (Cemen et al., 1985). In the Emigrant Canyon-Towne Pass area on the northwestern flank of the Panamint Mountains, Late Miocene through Pliocene conglomerates and **basaltic** to **rhyolitic** lava flows of the Nova Formation defines the Nova Basin (Fig. 3A). The Nova Formation has been deposited above a low-angle normal fault called the Emigrant detachment.

Basaltic. A compositional term used to describe magmas or volcanic rocks that consist of between 50-55% silica. Basaltic rocks are typically dark gray to black in color. Of or like the rock basalt. Generally, rocks with 55-60% silica are called basaltic andesites.

Rhyolitic. A compositional term used to describe magmas or volcanic rocks that consist of greater than 70% silica. Rhyolitic rocks tend to be light in color. Of or like the rock rhyolite.

Of the Cenozoic basinal deposits exposed within Death Valley National Park, those exposed within Furnace Creek Wash are the most obvious and scenic. These rocks define the Furnace Creek Basin, an accumulation of sedimentary units and basaltic to rhyolitic lava flows. These were deposited in Middle Miocene through Pliocene time on crust that lies between a major fault in the Furnace Creek fault zone (McAllister, 1970, 1971, 1973; Cemen et al. 1985) and the Badwater Turtleback fault (Wright et al., 1999; Miller, 1999). The crust there has moved downward, as well as laterally, to form the northeast margin of the elongate, trough-like Furnace Creek basin. The basinal deposits are about 3600 m (12,000 feet) in maximum estimated thickness in the northwestern part of the basin, but are much thinner in the southeastern part.

The panorama of the Furnace Creek Basin deposits viewed from Zabriskie Point is among the most photographed in the entire National Park system. These rocks are folded into a broad but complex **syncline**. Beginning on the west side of the Black Mountains, the Middle and Late Miocene Artist Drive Formation dips northeastward. The successively overlying Late Miocene to Pliocene Furnace Creek Formation and Pliocene Funeral Formation also dip northeastward and coincide with Furnace Creek Wash. There, conglomerates of the Furnace Creek Formation reflect deposition on alluvial fans while thinly bedded mudstones and sandstones reflect deposition in lakes (Photo 21A). Farther towards the Funeral Range, these rocks generally dip back towards the southwest.

Photo 21A. Furnace Creek Formation at Zabriskie Point, view looking southward. The conglomerate in the foreground accumulated in alluvial fans whereas the light-colored, fine-grained rocks in the middle ground accumulated in lakes.

In contrast, the interlayered conglomerates and volcanic rocks of the Nova basin are more disordered and thus less completely displayed than those of the Furnace Creek basin. This is largely because the basin was fragmented concurrently with the deposition of the Nova Formation and with movement on the underlying Emigrant detachment fault (Hodges et al., 1989).

Syncline. A fold in layered rock in which the youngest rock lies in the core. The beds on either side of a syncline generally tilt towards from the core.

Photo 21B. East-dipping Nova Formation west of Towne Pass.

The Nova Formation is estimated to be more than 1800 m (6000 feet) thick. It is exposed along State Highway 190 both east and west of Towne Pass (Photo 21B). The upper part of the Nova is well exposed in Emigrant Canyon on both sides of the road that connects State Highway 190 with Harrisburg Flat and Wildrose Canyon. The conglomerates there, and elsewhere in the formation, are derived from the Proterozoic formations exposed in the higher parts of the Panamint Mountains.

STOP AND LOOK

Other sedimentary basins that pre-date modern Death Valley include the Ubehebe basin in the northern part of the park (Snow and Lux, 1999), and the Copper Canyon basin near the southern part of the park (Knott, 1999, Fig. 3A). The Copper Canyon basin lies in the hangingwall of the Copper Canyon turtleback and has been folded into a series of northwest-trending **anticlines** and synclines (Drewes, 1963; Otton, 1976; Photo 21C).

Anticline. A fold in layered rock in which the oldest rock lies in the core. The beds on either side of an anticline generally tilt away from the core.

The Central Death Valley Plutonic-Volcanic Field

The Black Mountains and Greenwater Range and the adjoining Furnace Creek Wash, being underlain mostly by Cenozoic igneous and sedimentary rocks, present a varicolored patchwork-like landscape. This landscape differs markedly from the striped forms of the surrounding ranges composed of the uniformly layered Proterozoic and Paleozoic formations. The difference is equally visible on the geologic map (Photo 22, Fig. 3A).

Photo 21C. Lake beds of the Copper Canyon Formation. These rocks grade laterally into alluvial fan deposits.

The history of Cenozoic igneous activity within the central Death Valley region began between 12 and 11 million years ago (Wright et al., 1991). Within that interval the composite pluton of the Willow Spring **gabbro-diorite** was intruded at the site of the Black Mountains. Dacitic lava flows of that age are also exposed in the Greenwater Range and near the Resting Spring Range, east of the park. From 11 to 10 million years ago, felsic magmas crystallized both as shallow plutons and as lava flows. The felsic plutons are distributed throughout the central Death Valley area, but most abundantly in the Greenwater Range; the lava flows of that age are especially well exposed in the vicinity of Sheephead Mountain in the southern part of the field.

Gabbro. A compositional term used to describe intrusive igneous rocks that consist of between 50-60% silica. Gabbros are chemically equivalent to basalts. They are typically dark in color and contain no quartz.

Diorite. A compositional term used to describe intrusive igneous rocks that consist of between 60-70% silica. Diorites are chemically equivalent to andesites.

The post-10 million year history of the Central Death Valley plutonic-volcanic field is recorded primarily in extrusive volcanic rocks that eventually covered almost the entire area now occupied by the Black Mountains and Greenwater Range and, in the later stages, much of the adjacent Furnace Creek Basin. These rock units range widely from rhyolitic through andesitic to basaltic in composition (Greene, 1997). Rhyolitic lava flows and associated air-fall tuffs, comprising the Shoshone Volcanics, the volcanic rocks at Brown Peak, and the Greenwater Volcanics, are the most abundant. Most of the ash-flow tuff is confined to a single formation, the Rhodes Tuff which is about 9.2 million years old. An outstanding exposure of similar welded tuff of that age exists outside the national park on highway 178, approximately 6.5 km (4 miles) east of Shoshone (Troxel and Heydari, 1982).

Overlying the Rhodes Tuff are the Shoshone Volcanics, which are about 8 million years old. **Felsic** volcanism apparently terminated 5 to 6 million years ago following the emplacement of the Greenwater Volcanics. The basaltic and andesitic flows occupy various positions in the volcanic pile. The available radiometrically determined ages suggest a clustering in intervals of 9 to 10, 7 to 8 and 4 to 5 million years.

Felsic. A term used to describe igneous rocks that contain abundant feldspar.

Photo 22. View northwards along the crest of the northern Black Mountains. The brightly colored rocks are late Cenozoic volcanic rocks of mostly the Artist Drive Formation.

Amount and Style of Crustal Extension in Death Valley

A great deal of scientific controversy concerns the magnitude and style of extension in Death Valley, especially with respect to the Black Mountains. As shown on the geologic map (Fig. 3A) and discussed in the preceding section, the northern Black Mountains are nearly devoid of the Late Proterozoic and Paleozoic sedimentary rock that underlies the surrounding ranges. Instead, the northern Black Mountains consist primarily of Tertiary intrusive, volcanic, and sedimentary rock that either intrudes, overlies, or is faulted against basement rock.

One model for extension in Death Valley proposes that the older sedimentary rock was removed along a shallow-crustal, low-angle normal fault system that extended the entire region by about 80 km (Stewart, 1983; Wernicke et al., 1988; Holm et al., 1992; Snow and Wernicke, 2000). This model is generally referred to as the "rolling hinge model," because it requires the low-angle fault to steepen at the locus of major faulting (the "hinge"); the hinge migrates, or rolls, westward through time (fig. 9A). According to this model, the fault system is now exposed at the turtlebacks and the Amargosa Chaos, described in the previous chapter. It had to be most

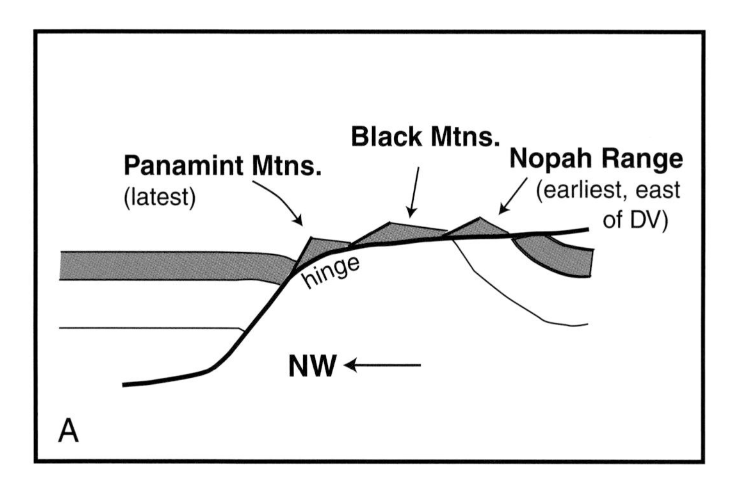

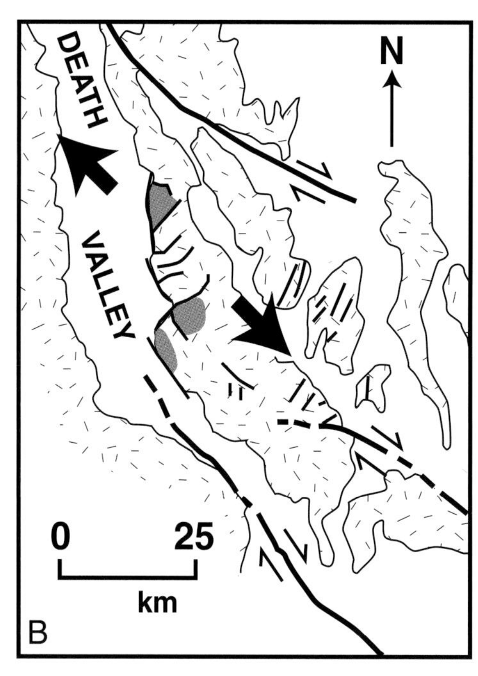

Figure 9. Models for late Tertiary extension in Death Valley, from Miller, 2001. 7A, Rolling Hinge model, in which active extension occurs at the hinge or bends in the detachment fault. As extension migrates to the northwest, so does the hinge. 2B, Pull-apart model, in which extension is driven by transfer of strain between terminations of large strike-slip faults. The two most likely faults for this scenario are the Furnace Creek Fault Zone (FC) and the Sheephead Fault (SF), although Serpa and Pavlis (1996) suggested that the Garlock Fault may also play a role. By this model, the Black Mountains have been extended significantly more than the surrounding ranges. Turtlebacks are shown as gray shaded areas.

active after about 8 Ma, to detach the Shoshone Volcanics in the Amargosa Chaos area, and to account for the cooling of the Mormon Point and Copper Canyon turtlebacks (Holm et al., 1992).

By contrast, the other model, called the "pull-apart" model, holds that the older sedimentary rock was removed along a series of separate, deeply rooted fault zones (Fig. 9B; Wright et al., 1991; Serpa and Pavlis, 1996; Wright et al., 1999). The pull-apart model therefore requires each turtleback and Amargosa Chaos to be separate features. Like the pull-apart model for modern extension, this model calls for extension to be driven largely by strike-slip. As a consequence, this model suggests that the Black Mountains have extended far more than the surrounding ranges, and in so doing, have made space for its voluminous Tertiary magmatism. Distinguishing between these two models is important because each calls on entirely different processes to extend the earth's crust. Moreover, because Death Valley is so well-known for crustal extension, the prevailing model in Death Valley naturally influences perspectives on crustal extension elsewhere.

In addition to the turtlebacks and the Amargosa Chaos, the Furnace Creek Basin plays a key role in testing these models. It contains a thick sequence of sedimentary and volcanic rock that is unbroken by any through-going detachment faults. The Rolling Hinge model therefore requires that the entire basin overlies the regional detachment fault and so was transported some 10s of km northwestward to its present location. In contrast, the pull-apart model requires that the basin formed in-place.

We favor the pull-apart model for a variety of reasons. It best explains the localization of voluminous magmatism within the distinctly rhombic-shaped region of the central and northern Black Mountains. In turn, this region is bordered by the Furnace Creek and Sheephead fault zones, along which right-lateral slip would naturally pull the area apart (Fig. 9B). Moreover, several observations of the Furnace Creek Basin, Amargosa Chaos, and the turtlebacks conflict with the rolling hinge model. At two localities in the northern Black Mountains, rocks of the Furnace Creek Basin rest depositionally on Paleozoic rock and are not faulted (mile 1.8 of Dantes View road guide; mile 66.2 of Black Mountains road guide). Additionally, Miller and Prave (2002) found that, prior to 14 Ma, faulting at the Badwater turtleback likely influenced sedimentation in the Furnace Creek Basin. If true, then the basin must have formed in approximately its present location.

In the Amargosa Chaos, Wright and Troxel (1984) mapped numerous depositional contacts between the Proterozoic sedimentary rock and the crystalline basement, to show that the Amargosa fault is not a single, through-going detachment fault, as required by the Rolling Hinge model. Miller (2002) limited the offset of one section of the fault to only 1 km. Additionally, most activity on the Amargosa fault must have been before 10 Ma, according to the age of a rhyolitic dike that intrudes the fault (Miller and Friedman, 2003) and new mapping and structural analysis by Topping (2003). These ages are long before the timing of major extension as required by the Rolling Hinge model.

At the turtlebacks, Miller and Pavlis (2004) described evidence that most of the extension along the ductile shear zones also occurred before 10 Ma, before that suggested by the Rolling Hinge model. This timing of extension was supported by Miller and Friedman (2003) who dated a dike that cut the shear zone at Mormon Point at 9.5 Ma. Furthermore, it is unlikely that the Panamint Mountains, for example, originated above and east of the Black Mountains as required by the rolling hinge model, because the Panamint Mountains contain Mesozoic intrusions that are not present in, or east of, the Black Mountains (Miller, 2002).

Chapter Three

Geologic History

Double rainbow and central Funeral Mountains.

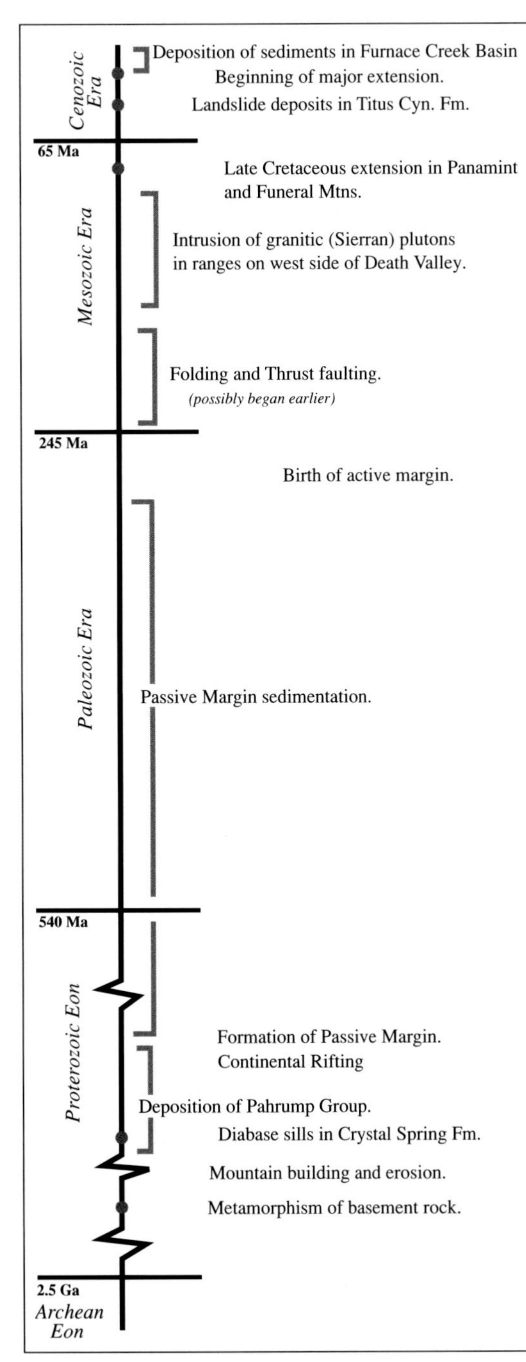

Figure 10. Simplified timeline of Death Valley's principal geologic events, from Miller, 2001. Gray bars generally denote approximate time spans of deposition, magmatism, or deformation. Arrows without a corresponding dot on the timeline mark timing of events relative to better constrained older and younger events.

By looking at the rock record of Death Valley, we can reconstruct the geologic history that preceded the formation of the present basins and ranges of the region. This procedure also involves a restoration of the displacements on the major faults that bound the present ranges. It is not a simple task, and markedly different reconstructions have been proposed. But geologists concerned with deciphering the earlier events generally agree that they include the ones shown in Table 1 and Figure 10. These primary events are described in the summary that follows.

The Ancient Basement and Prolonged Erosion

The oldest rocks in the Death Valley region are those of an Early Proterozoic crust that served as a basement for the thick accumulations of younger Proterozoic and Paleozoic rocks underlying most of the ranges of Death Valley. This basement consists of a complex of metamorphosed sedimentary and **igneous** rocks that are characterized by abundant quartz and feldspar. It is commonly called a **crystalline complex**. It can be recognized from a distance by its somber gray color and nearly featureless exposures. The complex underlies each of the turtleback surfaces along the Black Mountains front. It also is exposed in the southern part of the Black Mountains and in the Talc and Ibex Hills still farther to the south. A relatively small exposure of the basement also exists in the northern Funeral Range.

Igneous. A type of rock that formed by the cooling and crystallization of magma (liquid rock).

Crystalline complex. A body of metamorphic and igneous rock that forms the oldest and originally deepest part of a region.

The metamorphism of the basement has been dated by radiometric methods at about 1.7 billion years (Wasserburg et al., 1959; DeWitt et al., 1984). This is also the age of similar complexes exposed elsewhere in the southwestern United States. The complex contains belts of pegmatite dikes and widely distributed bodies of granite. A body of this ancient granite in the Panamint Mountains has been dated radiometrically at about 1.4 billion years. Except for the intrusion of the granite, little is known of the geologic history of the Death Valley region during the 500 million year interval between the metamorphism of the basement complex at 1.7 billion years, and the deposition of the lowest beds of overlying Pahrump Group at about 1.2 billion years. Clearly, however, this period contained at least one episode of uplift and erosion to expose the crystalline rock at the surface. Therefore, before the basal beds of the overlying Pahrump Group were laid down, large volumes of **detritus** must have been removed and then deposited.

Detritus. Erosional debris.

The Pahrump Group: Proterozoic Basins and Uplands

The sedimentary rocks of the Pahrump Group are characteristically several thousands of feet thick (see Basics About Measuring the Thickness of Sedimentary Rocks). The Pahrump Group is composed, in upward succession, of the Crystal Spring Formation, the Beck Spring Dolomite, and the Kingston Peak Formation. These formations underlie much of the west side of the Panamint Mountains. In the northern Funeral Mountains they occupy the northern part of the lower plate of the Boundary Canyon fault. They also are exposed at numerous localities in a belt that extends eastward from the southern Panamints to the east side of the Kingston Range. In the Funeral Mountains and in the central and northern part of the Panamint Mountains rocks of the Pahrump Group have been highly metamorphosed. Elsewhere they are little changed from their original state. Within the park boundaries, the most accessible exposures lie along and east of the front of the Black Mountains to the east and southeast of the Ashford Mill site (Wright and Troxel, 1984). All of the Pahrump units described below, as well as the overlying Noonday Dolomite, Ibex Formation, Johnnie Formation, and Stirling Quartzite, can be inspected in a half-day traverse of that locality.

The Pahrump Group chronicles a succession of events beginning more than a billion years ago when, in the region in and about Death Valley National Park, uplands of the crystalline complex rose rapidly above a shallow sea. The uplands were then dissected and partly buried in eroded sediment ranging from arkosic conglomeratic to muddy debris. These strata form the lower part of the Crystal Spring Formation (Wright et al., 1976; Roberts, 1976).

The entire seascape was then blanketed by a vast bank of dolomite and limestone growing in shallow water and covered by a laterally continuous **microbial mat**. This deposit forms the middle part of the Crystal Spring Formation. Stromatalites, which form by growth of algal mats in quiet water environments, are abundantly preserved in it. The mats were destroyed by the influx of fine-grained detritus preserved in the siltstones and sandstones in the upper part of the Crystal Spring. This event was followed by

Microbial mat. Literally, a mat of microbes. Such mats entrap fine-grained carbonate sediment that through time forms a stack of thin wavy layers of limestone. Their fossil forms are called stromatalites.

Basics About...

Measuring the Thickness of Sedimentary Rocks

Death Valley contains thousands of feet of Proterozoic sedimentary rock, tens of thousands of feet of Paleozoic sedimentary rock, and stratigraphically above that, several thousand feet of Cenozoic sedimentary rock. As thickness of a sedimentary unit is always measured perpendicular to bedding, it may seem impossible to measure, let alone see, such great accumulations of rock. However, we can see and measure these great thicknesses because the rocks are tilted as in the photograph below. In this way, one can virtually "walk into the past" by travelling in a direction opposite the inclination of the beds.

Photo 4-1. Pyramid Peak in the Funeral Range, consists of tilted Paleozoic sedimentary rock. By travelling in the direction of the arrow, one can view several thousand feet of increasingly older strata.

the intrusion of **diabase** sills at about 1 billion years ago, above and below the carbonate body (Photo 23; Heaman and Grotzinger, 1992). The lowest of the sills extended over hundreds of square miles of this ancient terrane and caused the basal carbonate strata to alter to bodies of commercial talc (see Basics About Mining in the Death Valley Area).

After Crystal Spring deposition, the environment reverted to that of another microbe-blanketed carbonate bank, now evidenced in the thinly laminated beds of the Beck Spring Dolomite. These and other Proterozoic microbial mats survived on such an extensive scale largely because animals that feed on algae had not yet evolved.

Diabase. A term used to describe a rock that is basaltic in composition, but crystallized at a shallow level in the crust rather than erupted on the surface and cooled as a lava flow.

Again the crust broke into blocks that formed islands in the Proterozoic sea. The Crystal Spring and Beck Spring rocks were exposed to erosion. Consequently, thick deposits of conglomerates containing abundant clasts derived from these earlier units, along with finer-grained clastic sediments, accumulated in basins between the new or rejuvenated high areas. These strata comprise the Kingston Peak Formation. Some of the conglomerates qualify as diamictites in that they contain a wide range of particle sizes set in a muddy matrix. Many clasts are boulder-sized or larger. Diamictites thus resemble glacial till. Other parts of the Kingston Peak consist of thinly laminated sandstone and siltstone in which clasts of boulder size and larger are embedded. For obvious reasons these are called "dropstones." The combination of the two textures provides evidence for a glacio-marine environment. Many geologists associate these deposits with a wave of glaciation that swept over the crust of North America 600 to 800 million years ago (eg. Hoffman and Schrag, 2000). Basaltic lava flows, locally present in this upper part of the Pahrump Group, mark a renewal of mafic magmatism.

Photo 23. Diabase sill (d) and arkosic sandstone (a) near the base of the Crystal Spring Formation. Here, Crystal Spring Formation lies depositionally on the basement. View is to the southeast, from near the top of Ashford Peak.

The Noonday Dolomite, Ibex Formation and Johnnie Formation: Foreshadowing the Environment of a Passive Continental Margin

The Noonday Dolomite, its deeper water equivalent called the Ibex Formation, and the overlying Johnnie Formation, directly overlie the Pahrump

Passive Continental margin. A continental margin, such as the east coast of North America, that does not border a convergent plate margin.

Group. They underlie a thick succession of uppermost Proterozoic and Paleozoic strata that has been long viewed as deposited along an evolving and **passive continental margin**. Therefore, the deposition of the three formations just preceded the splitting of a pre-existing continent into two continents separated by a widening expanse of oceanic crust.

The Noonday Dolomite, which is ordinarily about 300 m (1000 feet) thick, is yet another algal carbonate unit. Being a cliff-former and colored a distinctive pale yellowish gray, its exposures are particularly easy to recognize even from miles away. The Noonday and its lateral transition to the Ibex Formation are well-exposed above the Pahrump Group at the locality east of the Ashford Mill Site (Wright and Troxel, 1984).

From the Panamint Mountains as far east as the Kingston Range, the contact between the Noonday Dolomite and the underlying Pahrump is an angular unconformity that truncates progressively older units of the Pahrump Group northward. In its most northerly exposures along this belt, the Noonday Dolomite overlies the ancient basement complex. These relationships define a high area in the ancient topography from which much or all of the Pahrump Group was eroded before deposition of the Noonday (Fig. 10). At more southerly localities, either the Noonday or the Noonday-correlative Ibex Formation lie concordantly upon the Kingston Peak Formation. In the lateral transition from the Noonday Dolomite to the Ibex Formation, the shallow water, algae-related carbonate strata abruptly give way to a succession composed mostly of deeper water, thinly bedded siltstone and limestone (Williams et al., 1976; Wright et al., 1976; Wright and Prave, 1992).

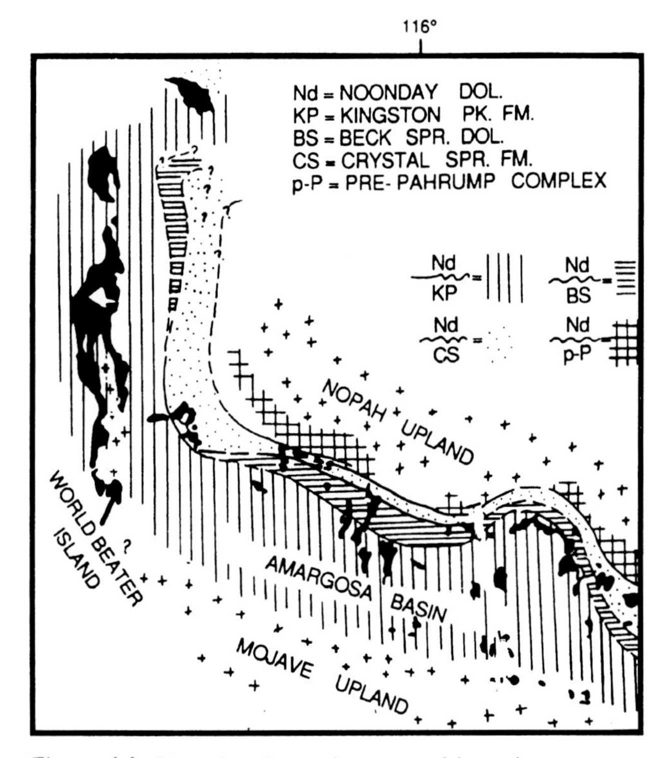

Figure 11. Map showing palogeographic and paleogeologic features of the Death Valley region just before the Noonday Dolomite was deposited. Black areas indicate the present exposures of the Pahrump Group. These extend from the Panamint Mountains on the west to the Kingston Range on the east and from the Silurian Hills on the southeast to the northern Funeral Mountains on the north. Linear and dotted patterns indicate the rock units that underlie the Noonday. Crosses show high areas from which debris was shed in Kingston peak time. Original pattern has been distorted by Mesozoic compression and Cenozoic extension. From Wright and Prave, 1992.

The Stirling Quartzite, Wood Canyon Formation, and Zabriskie Quartzite: The Clastic Wedge of a Developing Continental Margin

These Late Proterozoic to Early Paleozoic formations (Table 1), are widely exposed in the Death Valley region to average about 1800 m (6000 feet) in combined thickness. As they consist mostly of well-cemented sandstones and subordinate conglomerates, they are more resistant than the underlying, varicolored and shaly Johnnie Formation and the overlying, also varicolored Carrara Formation. The Wood Canyon Formation is particularly accessible west and north of the highway in the vicinity of Hells Gate between the Funeral and Grapevine Mountains. All three formations are well-exposed on the north face of Tucki Mountain at the north end of the Panamint Mountains.

By analogy with sedimentary wedges that have accumulated along present-day continental margins, geologists view this succession of clastic rocks as deposited during the rifting stage of an early margin. They reason that, at that earlier time, the ancient quartz- and feldspar-rich basement complex was split in two and was exposed along the edges of the two developing continents. Thus the Stirling Quartzite, Wood Canyon Formation, and Zabriskie Quartzite are interpreted as debris eroded from the edge of the more easterly continent and deposited as a wedge on the newly rifted crust (Fig. 12; Diehl, 1976).

These formations also provide the Death Valley region's earliest evidence of complex life forms. They contain the remains of metazoan life forms, particularly trilobites, and other features such as tracks and burrows, left by these creatures which rapidly appeared and populated the early oceans (Photo 24). In recent years fossils of the Late Proterozoic Ediacara fauna have been found in the Wood Canyon Formation (Corsetti and Hagadorn, 2000; Horodyski et al., 1994).

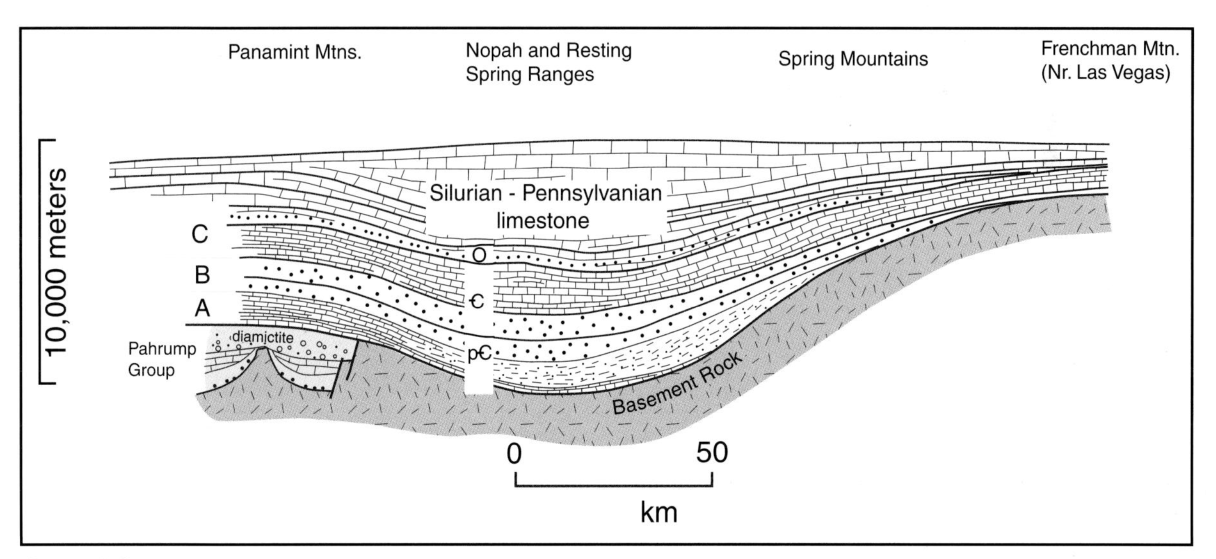

Figure 12. Stratigraphic cross section showing the Proterozoic and Paleozoic sedimentary succession exposed in the region between Las Vegas Valley and Panamint Valley. Brick pattern represents carbonate rock; dot and dashed patterns represent clastic rock; gray stippled pattern represents basement rock. The thickening and the upward transition from quartzitic to carbonate strata characterize a passive continental margin. Simplified from Wright et al., 1981.

Photo 24. Trilobite fragments in the Wood Canyon Formation. The Precambrian-Cambrian boundary is marked by the appearance in the rock record of hard-bodied fossils such as these trilobites. Their absence in the lower part of the Wood Canyon Formation, but presence in its upper part, suggests that the Wood Canyon Formation was deposited during a period of time that spanned the Precambrian-Cambrian boundary.

The Paleozoic and Early Mesozoic Carbonate Shelf

The rest of the Paleozoic section in the Death Valley region is dominated by dolomites and limestones that reach about 6100 m (20,000 feet) in thickness (Fig. 12; Photos 20, 25). These formations constitute the sedimentary record of a long lived and slowly subsiding continental shelf. Here they were deposited, with little evidence of disturbance, above the wedge of clastic sediments deposited during the preceding stage of rifting. The carbonate sedimentation was interrupted by numerous periods of emergence and, in Middle Ordovician time, by the deposition of the Eureka Quartzite. The Eureka is part of a sheet of quartz-rich sand that spread across much of the width of that Paleozoic continent. Carbonate rocks continued to be deposited into Triassic time. But then the sea withdrew and did not return. The oceanic crust, which was theoretically forming as the continental shelf evolved, lay many miles to the west of the site of Death Valley.

Photo 25. Paleozoic carbonate in the Funeral Range. View from near top of Pyramid Peak, looking eastward.

Where exposed within the park boundaries, the carbonate succession is about 6100 m (20,000 feet) thick. It underlies large parts of the Panamint, Cottonwood, Grapevine and Funeral Mountains and is particularly visible in the southern part of the Funeral Mountains. There the Eureka Quartzite can be recognized from a distance as a single, nearly white band, repeated several times by normal faulting. It is underlain by the Pogonip Group, colored various shades of gray, and overlain by the nearly black Ely Springs Dolomite. In Death Valley National Park the Triassic strata are preserved only in the Butte Valley area at the southern end of the Panamint Mountains.

The Closing of the Paleozoic Ocean in Mesozoic Time

At numerous localities within Death Valley National Park we observe that the Proterozoic and Paleozoic rocks have been faulted so that older formations override younger formations. These faults, called thrust faults, record severe contraction of the crust. Together with associated folds, the thrust faulting marks the end of the passive continental margin and records a foreshortening of the former continental shelf. Some of the more accessible

Photo 26A. Tightly folded limestone of the Devonian Lost Burro formation. View is to the west towards Saline Valley.

thrusts are the Schwaub Peak (Wright and Troxel, 1993) and Clery thrusts (Cemen et al. 1985) in the southern part of the Funeral Mountains. In northern Death Valley, the Last Chance Thrust is readily accessible from the road to Eureka Valley. Two spectacular folds in the Grapevine Mountains, the Corkscrew Peak Syncline and the Titus Canyon Anticline, are visible from just west of Daylight Pass and the Titus Canyon road, respectively (Photos 26A, B). However, some researchers interpret their geometries to have resulted, at least partially, from Tertiary extension (Reynolds et al., 1996). Visitors who travel down the Titus Canyon Road pass through the core of the Titus Canyon Anticline approximately 1.6 km (1 mile) below Klare Spring. Beyond that point, all the rocks in the canyon, including those that appear to be nearly horizontal, are overturned.

Death Valley also hosts several Mesozoic plutons, of both Jurassic and Cretaceous ages, shown in green on Figure 3A. They typically lie close to the park's western boundary in the Owlshead Mountains, Panamint Mountains, and the southern end of the Cottonwood Mountains (Photo 11A). Most of these plutons are accessible by unimproved roads.

Together with zones of volcanic rocks, Mesozoic plutons characterize a belt that extends beyond the park area for the full length of North America. Called the "Cordilleran magmatic arc," this belt includes the Sierra Nevada batholith to the west, the Peninsular Ranges batholith to the south, and the Coast Plutonic Complex as far north as British Columbia. These granitic bodies all formed within roughly the same geologic setting, where an oceanic lithospheric plate dove beneath a continental plate. Today, a similar environment exists in the Pacific Northwest, where the oceanic Juan de Fuca plate dives beneath North America. The resulting magmatism fuels eruptions of the Cascade volcanoes.

Photo 26B. Corkscrew Peak recumbent Syncline. Corkscrew Peak gains its peculiar shape from the folding of Paleozoic rock near its summit. In this straight-on view looking nearly due east, one can see a continuous bed of Bonanza King Formation that folds completely over itself on the right-hand side of the photo. Although faulted in many places, this fold is continuous with the Titus Canyon Anticline of photo 26A.

The Missing Sedimentary Record and Beginnings of the Basin and Range Event

No sedimentary rocks of Jurassic through Eocene age have been found in the area of Death Valley National Park, although volcanic rocks of probable Jurassic age are exposed in the Butte Valley area of the southern Panamint Mountains. In the national park and the surrounding region, this was apparently a time of uplift and erosion concurrent with and following the thrust faulting and intrusion of the Mesozoic plutons. During this interval, much of the succession of Proterozoic and Paleozoic sedimentary rocks that once covered the Death Valley region was eroded away. In fact, the erosional denudation probably continued over much of the Death Valley region until the inception of the basins and ranges that form the present landscape.

Some research suggests that during the latest Cretaceous, Death Valley underwent a period of extension after thrust faulting. In the Panamint Mountains, Hodges and Walker (1992) inferred an episode of Late Cretaceous to early Tertiary extensional uplift from the timing of cooling of metamorphic rocks. In the northern Funeral Range, Applegate et al., (1992) proposed a period of uplift associated with slip on normal faults between 70 and 72 Ma.

The Oligocene and Early Miocene Flood Plain and the Bordering Upland

The Oligocene and early Miocene formations of the Cottonwood, Funeral and Grapevine Mountains were deposited on a much more subdued landscape than the one we observe today. This conclusion stems from the observation that the contact is characteristically devoid of the major irregularities that would accompany the infilling of an irregular topography. We also note that basal Tertiary beds consistently dip almost as steeply as the underlying Proterozoic and Paleozoic formations of the present ranges. Thus, when rotated back to their original, nearly horizontal positions, these beds are seen as deposited on a broad surface of low relief drained by laterally migrating streams and rivers.

STOP AND LOOK

The oldest Tertiary unit, the Oligocene Titus Canyon Formation, consists of sedimentary breccias, conglomerates, sandstones and mudstones (Photo 27). The breccias, exposed near Red Pass on the Titus Canyon Road, are probably landslide deposits. They suggest the presence of adjacent highlands during the Oligocene (Reynolds, 1969, 1976; Saylor and Hodges, 1991). The conglomerates, however, suggest more distant sources. Their clasts tend to be exceptionally well-rounded and polished. They also contain granitic rock that was probably derived from a westerly source area in which Mesozoic plutons were exposed. Good exposures of these conglomerates exist in road cuts near Daylight Pass and along the Titus Canyon road.

Photo 27. Oligocene Titus Canyon Formation (Tot) and overlying Late Tertiary volcanic rock (Tv) at Red pass in the Grapevine Mountains. Arrow points to contact.

In a more southerly area of the national park that includes most of the Panamint Mountains and the Owlshead and Black Mountains, Tertiary rocks older than 14 million years are apparently absent, and the pre-Basin and Range surface is thus obscured. If so, the missing rock units of earlier Tertiary age have either been eroded away or were never deposited. In this area, the 14 million-year-old and younger Tertiary sedimentary rocks, are part of the evolving Basin and Range terrain.

Late Cenozoic Extension

Most workers in Death Valley agree that the most recent period of extension in Death Valley began as far back as 16 Ma (Wright et al., 1999). We can track the extensional history of Death Valley in detail by studying various sedimentary deposits such as those of the Furnace Creek and Nova Basins or specific features within some of the fault zones (Miller, 1999).

Death Valley itself, however, did not form until slip on the Black Mountains Fault Zone began, which was almost certainly after 4 Ma. Prior to

4 Ma, the area near Ryan in upper Furnace Creek Wash was strongly deformed but a prominent angular unconformity beneath flat-lying, 4 Ma basalt marks the cessation of deformation (Photo 28). Perhaps more significantly, the earliest deposits that seem to indicate uplift of the Black Mountains are post-3 Ma deposits of the Funeral Formation.

Photo 28. View of Ryan mining camp and Ryan Mesa. Gently dipping Miocene Artist Drive Formation (Tma) is faulted against moderately dipping Upper Miocene-Pliocene Furnace Creek Formation (Tmf). Undeformed basalt flows of the Pliocene Funeral Formation (Tpf) overlies them.

Chapter Four

Death Valley Road Guides

Highway 178, looking east towards volcanic rocks of the Dublin Hills (middle ground) and Paleozoic rocks of the Nopah Range (background).

Introduction

Much of the spectacular geology of Death Valley is accessible or visible from the park roads, naturally lending the area outstanding field trips. A great deal of active research is conducted in Death Valley, providing opportunities to investigate geology in depth.

This chapter provides geology guides for each of the major roads in the park, including the road down Titus Canyon. Because many of these localities offer opportunities for advanced study, detailed information and additional references to scientific literature are provided.

Symbols

Throughout this road guide the following symbols will aid in traveling through the park.

This symbol indicates good opportunities for short hikes to see the geology up close. It also indicates the need for special care as even short hikes are potentially dangerous. Hikers should bring plenty of water and be familiar with the rigors of this beautiful, but harsh, desert landscape.

This symbol indicates that road conditions warrant the use of a 4-wheel drive vehicle for passage.

This symbol is a reminder that the National Park Service does not allow sample collecting within Death Valley National Park without a permit.

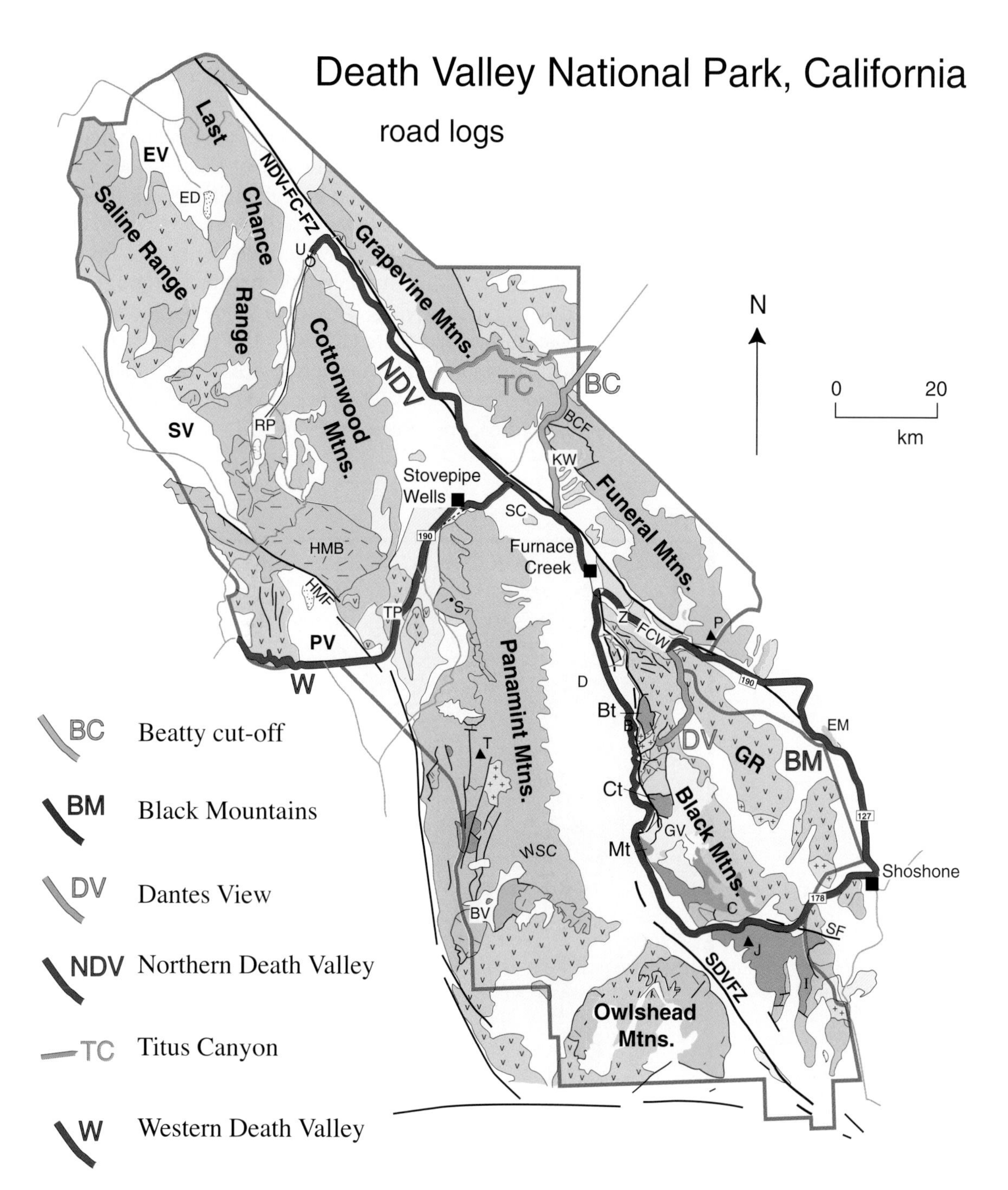

Figure 13. Road logs for Death Valley National Park. Different colored roads and accompanying abbreviations indicate different guides. Abbreviations (in black) are as follows: BCF: Boundary Canyon fault; Bt: Badwater turtleback; C: Chaos; Ct: Copper Canyon turtleback; D: Devil's Golf Course; ED: Eureka Dunes; EM: Eagle Mountain; EV: Eureka Valley; GR: Greenwater Range; GV: Gold Valley; HMB: Hunter Mountain batholith; HMF: Hunter Mountain fault; KW Keane Wonder fault; Mt: Mormon Point turtleback; NDV-FC-FZ: Northern Death Valley-Furnace Creek fault zone; PV: Panamint Valley; P: Pyramid Peak; RP: Racetrack Playa; SV: Saline Valley; SC: Salt Creek; SDVFZ: Southern Death Valley fault zone; SF: Sheephead fault; T: Telescope Peak; TP: Towne Pass; WSC: Warm Springs Canyon.

Road and Field Guide to the Northern Black Mountains

In a geological sense, the northern Black Mountains are the most unusual range in Death Valley because they consist primarily of Late Tertiary intrusive and volcanic rock as well as Precambrian basement. They lack the thick sequences of Late Proterozoic through Paleozoic sedimentary rock that characterize the other ranges. The northern Black Mountains are known for their dramatic fault-bounded west front, turtlebacks, badlands topography, Amargosa Chaos, and wineglass canyons. This field guide begins at the intersection of the Badwater Road and State Highway 190 in front of the Furnace Creek Inn and then circumnavigates the northern Black Mountains and Greenwater Range in a counterclockwise direction. Its route, as well as some recommended stopping places, are shown on Figure 14.

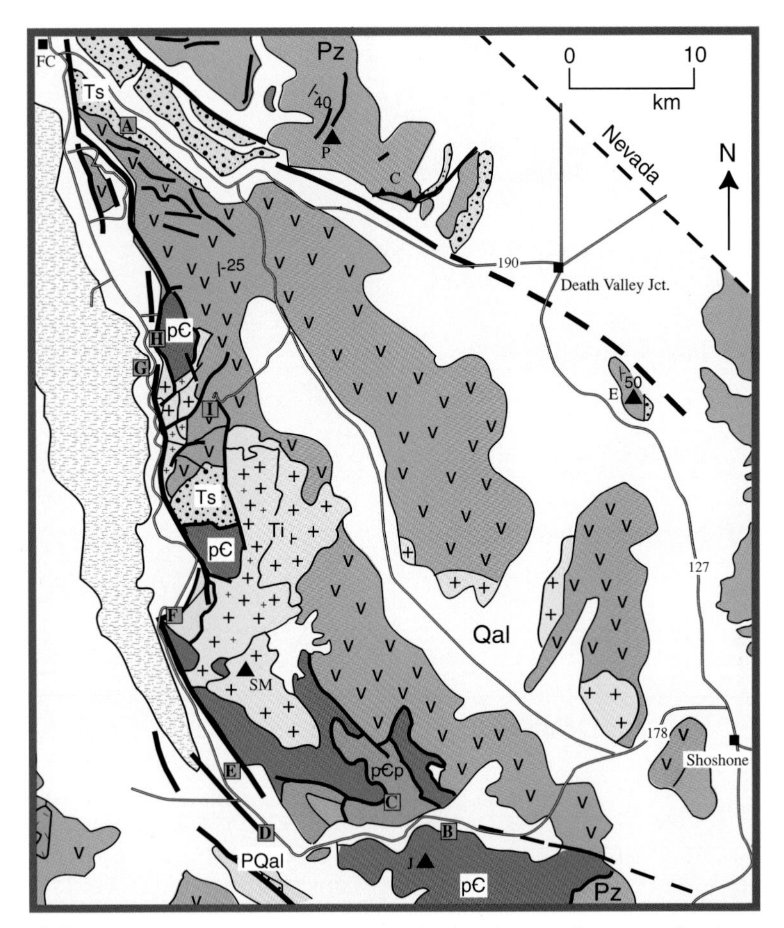

Figure 14. Map of northern Black Mountains. Refer to Figure 1A for legend. Field trip locations are signified by letters enclosed in squares. Abbreviations consist of C: Clery Thrust; E: Eagle Mountain; FC: Furnace Creek; J: Jubilee Mountain; P: Pyramid Peak; SM: Smith Mountain.

Mileage (Intersection of Highway 190 and Badwater Road)

[Reverse mileages are from intersection of Highway 127 & 178 north of Shoshone]

0.2 mi [53.6] Northeast tilted conglomerate of the Furnace Creek Formation Cemen et al. (1985) and Wright et al. (1999) show that the conglomerates of the Furnace Creek Formation are part of an alluvial fan complex that was derived from the west. The small building perched atop the cliffs is a teahouse, operated by the staff of the Furnace Creek Inn during the 1930's and 1940's.

0.6 mi [53.2] Large deposit of **travertine** on the left. Note that the travertine cuts across bedding in the underlying Furnace Creek Formation to indicate that it is a younger feature.

0.7 mi [53.1] Angular unconformity between tilted Furnace Creek Formation and overlying, near-horizontal, alluvial fan gravels. This unconformity shows that most deformation in Furnace Creek Wash had occurred prior to deposition of the gravels.

1.3 mi [52.5] Palm trees and other lush vegetation in Furnace Creek Wash mark small springs in the wash. Travertine Spring, which can produce water flows as great as 990 gallons/minute (Steinkampf and Werrell, 2001), lies a short distance up the alluvial fan to the northeast. Travertine Spring is the main source of freshwater in the Furnace Creek area.

2.0 mi [51.8] Echo Canyon road on left. This road provides good access to the Funeral Range, including the Inyo gold mine and starting points for hikes up Winters and Schwab Peaks. At the mountain front, approximately 4 miles away, a small dryfall requires 4WD and high clearance. Less than 1/4 mile up the road, a thrust fault can be seen cutting the alluvial fan gravels. This fault is part of the Echo Canyon thrust, described in detail by Klinger et al., 2001. (p. B75)

3.5 mi [50.3] Zabriskie Point, Location A on Fig. 14. Zabriskie Point (photo 21A, page 43), as well as 20 Mule Team Canyon (mile 4.6) provides outstanding views of badlands topography, formed in tan, white, and yellow-colored lakebeds of the Furnace Creek Formation. Badlands topography is discussed on p. 11. The prominent dark rock in the middle of the badlands is a basaltic intrusion, as mapped by McAllister (1970). The red-colored cliffs immediately north of Zabriskie Point consists of alluvial fan deposits, also of the Furnace Creek Fm. A trail, beginning on the north side of the parking lot, leads down into the badlands, and exposures of both the alluvial fan and lake deposits. Hikers on this trail can continue down either Golden Canyon or Gower Gulch to the valley floor.

On the southeast side of the parking lot, visitors can see how Furnace Creek Wash has been diverted into a narrow slot that empties into Gower Gulch. This diversion was a flood control effort, built in 1941. Before the diversion, Gower Gulch drained approximately 2 sq mile of land; it now drains approximately 170 sq miles. The diversion explains why the gravel that covers the floor of Gower Gulch consists largely of Paleozoic rock, rather than the adjacent soft siltstone and sandstone of the Furnace Creek Fm. Environmental effects of the diversion are documented by Troxel (1974).

4.6 mi [49.2] Entrance to Twenty Mule Team Canyon. This short drive (2WD accessible) winds through beautiful badlands of the Furnace Creek Formation. Locally, the lake deposits are cut by basaltic intrusions like the one at Zabriskie Point. The road also provides a good starting point for hikes into the Black Mountains.

The road continues through northeast-dipping outcrops of the Furnace Creek Formation.

5.4 mi [48.4] Road to Hole-in-the-Wall. Hole in the Wall is a prominent stream-carved gash in a ridge of steeply southwest-dipping alluvial fan deposits of the Furnace Creek Formation. The ridge forms the steep northeastern limb of the Furnace Creek Syncline, an asymmetrical fold that trends parallel to Furnace Creek Wash (p. 43). Just past the ridge, the Furnace Creek Fault zone places Paleozoic rock of the Funeral Mountains against the Furnace Creek Formation.

At this point, the road enters northeast-dipping exposures of the Funeral Formation. They are folded into a north-trending syncline, which is truncated on its southeastern side by the Cross-Valley fault of McAllister (1970).

6.2 mi [47.6] Exit of Twenty Mule Team Canyon road.

10.7 mi [43.1] Dantes View Road (see page 81 for roadlog to Dantes View)

11.9 mi [41.9] Vertical travertine-filled fractures in Funeral Formation.

13.2 mi [40.6] Folds and minor thrust faults of Furnace Creek Formation on the south side of the road.

16.8 mi [37.0] Elevation 3000' sign (ascending).

Between here and Death Valley Junction, the road follows the trace of the Furnace Creek Fault Zone and the edge of the Funeral Mountains (p. 39–40). Pyramid Peak, the tallest peak in this part of the range, is capped by a small pyramid of black Ordovician Ely Springs Dolomite, which overlies the white Ordovician Eureka Quartzite. At approximately mile 21, one can see light colored middle Tertiary rocks high in the Funeral Mountains. These rocks overlie a low-angle normal fault that was previously a thrust fault (the Clery Thrust) during the Mesozoic (Cemen and Wright, 1990). At the southeastern end of the range, the Tertiary sedimentary rocks unconformably overlie Mississippian limestone of the Perdido Formation (McAllister, 1976).

Photo 29. Amargosa River in flood. View is south from near Eagle Mountain.

28.6 mi [25.2] Intersection with CA Highway 127

28.8 mi [25.0] Death Valley Junction and the Amargosa Opera House. Death Valley Junction originated as a company town for the Pacific Coast Borax Company in 1923. The buildings are made of adobe and form an L-shape in plain view, with the largest building at the northeast end. This large building was the recreation hall.

In 1968, Marta Beckett, an artist and dancer from New York, reopened the hall as the Amargosa Opera House. She performed three times a week, even without an audience, and decorated the inside walls and ceiling with stunning murals. As of 2003, she still performs during the late fall, winter, and spring months. Thanks to her efforts, Death Valley Junction became listed on the National Register of Historic Places in 1981.

36.0 mi [17.8] Road turns east towards Eagle Mountain, a tilted fault-block of Cambrian limestone.

Several normal faults can be seen on the west side of the mountain.

39.1 mi [14.7] Road crosses the Amargosa River. Although a dry wash for most of the year, this part of the Amargosa River flows vigorously after major rain storms (Photo 29). The river continues southward past Shoshone and Tecopa to the Dumont dunes, where it turns to the west and then back to the north in the vicinity of Saratoga Springs. From there, the river flows northward into Death Valley. Near Tecopa, where it is fed continually by springs, the Amargosa River flows year-round.

Here, on the southwestern side of Eagle Mountain, one can see 15 Ma sedimentary rock (tan-colored material) that unconformably overlies the Cambrian limestone. This younger rock contains granitic clasts that were derived from the Hunter Mountain Batholith, approximately 80 km (50 miles) to the northwest. Niemi et al. (2001) argued that the presence of these clasts indicate the rocks were transported that far by slip on a low-angle detachment system. Other geologists, however, dispute this interpretation, and instead argue the clasts were transported by a river system. (Renik & Christie-Blick, 2004)

53.8 mi [0.0] Intersection of CA Highway 127 with CA Highway 178 (The Badwater Road). Turn right and re-set odometer or continue south one mile to the town of Shoshone. Two cafes in Shoshone, the Crowbar Café and C'est Si Bon, both offer good food. Also recommended is a visit to the Shoshone Museum.

Reset Odometer, 0.0

For the first couple miles, the road climbs an alluvial fan, with views of faulted Proterozoic and Paleozoic rocks of the Dublin Hills to the south. Beginning at about 2 miles, Tertiary volcanic rocks come into view on the left. In most places, the volcanic rocks are faulted against the sedimentary rocks, but locally they overlie them depositionally (Chesterman, 1973).

[Reverse mileage in brackets is from intersection of Badwater Road and Highway 190]

2.5 mi [67.4] Good view of Chocolate Sundae Mountain ahead to the right. The tan-colored rocks are part of the granitic Chocolate Sundae Mountain pluton, and the tilted multicolored rocks are part of the Shoshone Volcanics (Haefner, 1976). The Shoshone Volcanics consist of rhyolitic lava flows and tuffs, which here rest depositionally on the pluton. The volcanic rocks are only slightly younger than the pluton, suggesting that the pluton rose from shallow depths soon after it cooled. This rising caused the volcanics to tilt away from the pluton (Haefner and Troxel, 2002).

5.7 mi [64.2] Intersection (north) with Furnace Creek Road, otherwise known as the Greenwater Road. This gravel road provides access to the Greenwater Valley. It joins the Dantes View Road after about 28 miles. High clearance recommended.

7.0 mi [62.9] View back to the northeast of Chocolate Sundae Mountain. Here, one can see how the peak got its informal name: rocks from its basaltic cap spill off the top of the peak like chocolate sauce!

10.4 mi [59.5] Well exposed normal faults in volcanic rocks to the south.

10.6 mi [59.3] Salisbury Pass. There is ample room for parking here and access to the Shoshone Volcanics and the underlying Rhodes Tuff on both sides of the highway. The isolated blocks south of the highway are ash flow tuffs that depositionally overlie Paleozoic rock. A few outcrops of quartzite are exposed in low-relief areas between the blocks.

15.5 mi [54.4] Starting about here, and continuing until about mile 31, geology of the route is depicted in detail on the map by Wright and Troxel (1984) (Map 2, page 107).

16.4 mi [53.5] Outcrop of Proterozoic diabase sill in basement rock to the right.

17.6 mi [52.3] "Exclamation Rock," Location B on Figure 14 and Map Y. Pull off onto the wide shoulder to the right. The cliff on the south side of the road provides an unusually instructive example of the Virgin Spring phase of the Amargosa Chaos, mapped by Wright and Troxel, 1984. Although only about 230 feet high, the cliff displays fault-bounded slivers of the crystalline basement, Proterozoic diabase, Lower Crystal Spring Formation (lowest rock unit of the Pahrump Group), and the Noonday Dolomite. These rocks appear in proper stratigraphic order, but more than 1640 feet are missing along the fault zones. To identify the specific units, refer to Photo 30.

Photo 30. View of "Exclamation Rock" with fault contacts drawn to illustrate structure of Amargosa Chaos. The entire part of the upper Crystal Spring Formation is missing beneath the Lower Noonday Dolomite. Absence of Beck Spring Dolomite and Kingston Peak Formations, however, is likely because those units stratigraphically pinch-out in this vicinity.

20.1 mi [49.8] Jubilee Pass. There is parking for a few vehicles on the north side of the road just beyond the pass. To the south lies basement rock and Jubilee Mountain, an easy climb (approximately 1300 feet) for those who seek especially good views of southern Death Valley and the Owlshead Mountains. Much of the basement rock consists of augen gneiss that is locally cut by Late Proterozoic diabase. The basement rock is discussed in greater detail on p. 50–51.

20.5 mi [49.4] Turn-off for access to Virgin Spring Canyon. This gravel road leads approximately 1.3 mile to a small turn-out. From here, one can walk about 1/2 mile to the just past the entrance to Virgin Spring Canyon, Location C on Figure 14 and Map Y. Here, on the west wall, is a spectacular exposure of the Virgin Spring phase of the Chaos as shown in Fig. 7, page 38.

21.6 mi [48.3] Point of Rocks. The tilted sedimentary rocks on the south side of the road are part of a late Tertiary sedimentary basin mapped and described by Topping (1993) and Wright and Troxel (1984). These particular rocks contain distinctive clasts that indicate they were derived from the Panamint Mountains. On the west side of the outcrop, they are faulted against granitic rock avalanche deposits. The rock avalanche deposits, which exist on both sides of the road, can be identified by their distinctive cavernous weathering. In contrast to the well-layered sedimentary rocks, Topping (1993) argued that the rock avalanche deposits were derived from the Kingston Range, more than 30 miles to the southeast.

22.5 mi [47.4] Good view of the Owlshead Mountains to the southwest. They are capped by Late Tertiary basaltic lava flows overlying Mesozoic granitic rocks.

23.0 mi [46.9] Road passes a large hill of Beck Spring Dolomite (middle unit of the Pahrump Group).

23.7 mi [46.2] Pull-out to access Upper Kingston Peak Formation (upper unit of the Pahrump Group) and diamictite deposit beneath Noonday Dolomite. Park at west edge of large hill of Pahrump Group, immediately south of the road. Hike approximately 1/3 mile north-northeast to the first canyon that cuts into the range. Good exposures of the diamictite can be found along the northern wall of the canyon. These rocks are overlain by carbonate rocks of the Noonday Dolomite.

Photo 31. View of Pahrump Group and Noonday Dolomite from Ashford Mill. The lowest rocks exposed on the left side are crystalline basement; orange and red rocks overlying the basement are carbonate and clastic rocks of the Crystal Spring Formation; green rock within the Crystal Spring Formation are diabase sills. Beck Spring Dolomite appears as the light gray rock above the Crystal Spring Formation. On the right-hand side of the photograph, Kingston Peak Formation appears as the red-colored ridgetop overlying the Beck Spring Dolomite; it pinches out towards the west (left), where Noonday Dolomite appears as the tan unit overlying the Beck Spring Dolomite.

24.7 mi [45.2] Road to Baker, California joins from south. The Amargosa River occupies the bottom of the valley floor. It flows north from here to empty into Death Valley. The dirt and gravel road provides access to southernmost Death Valley National Park. It requires 4WD because it traverses many sandy stretches along the Amargosa River.

26.7 mi [43.2] Ashford Mill, pit toilet and Location D on Figure 14. Access road to Ashford Canyon (right). The road to Ashford Canyon requires high-clearance and 4WD.

North of the highway, there is a good view of the Pahrump Group (Photo 31). In general, the bright yellow-tan rocks near the base of the mountain are carbonate rocks of the Crystal Spring Formation, the bluish gray rocks are the Beck Springs Dolomite, and the dark-colored rocks are the Kingston Peak Formation. The overlying tan-colored rocks are the Noonday Dolomite. Intruding the Crystal Spring Formation are dark green-colored sills of 1.08 billion year old diabase (Heaman and Grotzinger, 1992; Photo 23). Pages 51, 53–54 discuss the Pahrump Group and overlying Noonday Dolomite in more detail.

28.0 mi [41.9] Lake Manly informational sign. The dark-colored hill behind the sign is Shoreline Butte (Photo 12), so named because of the many horizontal benches visible from the road. These benches originated as wave-cut shorelines along the edge of Lake Manly, which periodically filled the floor of Death Valley between 216 to 18 thousand years ago (Blackwelder, 1933; Machette et al., 2001). The highest bench at Shoreline Butte is at an elevation of about 300 feet, to indicate a maximum water depth for the lake of nearly 600 feet.

28.5 mi [41.4] West Side Road. A short drive down West Side Road leads to a cinder cone, offset right-laterally by the southern Death Valley fault zone (Photo 13B, page 25). The ridge of basalt along the north side of the highway appears to be a faulted lava flow.

30.6 mi [39.3] Location E on Figure 14. Pull off the highway at the crest of the hill for good views of the offset cinder cone to the southeast, and the range-frontal fault of the Black Mountains to the north. The range-frontal fault is expressed as the abrupt transition from the valley floor to the mountains. Note the Late Cenozoic fanglomerate that is faulted against the rangefront, but is uplifted with respect to the valley floor. Here, the most recent faulting appears to leave the Black Mountains range-frontal fault and step across the valley to the southern Death Valley fault zone (Photo 32). Several wineglass canyons can be seen at the rangefront.

33.7 mi [36.2] Road passes through several large debris flows. Notice the spring at the toe of the deposits. The sloping top of the Mormon Point turtleback (p. 7, Photo 5B) is becoming increasingly clear to the northwest. In general, the tan-colored rocks are carbonate rocks whereas the dark-colored rocks are part of the crystalline basement. Both rock types are highly deformed, as they exist in the footwall of this major detachment fault.

35.6 mi [34.3] Fault-controlled spring. Notice the many exposures of the Mormon Point turtleback fault in the small side canyons north of the highway.

36.9 mi [33.0] An especially good exposure of the Mormon Point turtleback fault. Park on the wide shoulder and walk into the narrow canyon. Hayman et al., (2003) argue that the Mormon Point and Badwater turtleback faults slipped at their present low angle.

37.5 mi [32.4] Sign for Mormon Point.

37.9 mi [32.0] Location F on Figure 14. Good pull-out to view the salt pan, Panamint Range, and features of the Mormon Point and Copper Canyon turtlebacks. See p. 6–7, 33 and Photos 5A, 5B for a more detailed discussion of the turtlebacks. A fault scarp cuts the alluvial fan on the east side of the road.

Mormon Point marks a boundary between the Badwater Basin to the north and the Mormon Point Basin to the south; they are likely separated by a right-lateral fault, coincident with the range

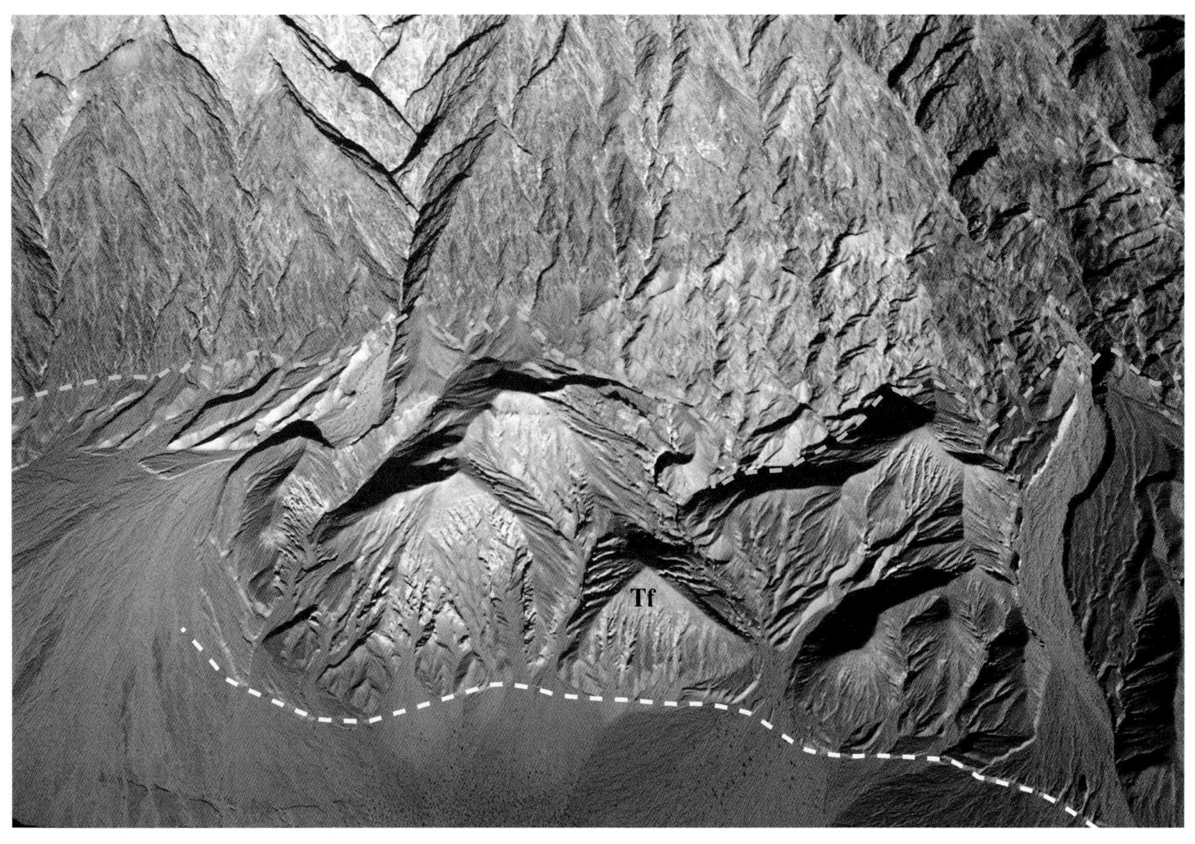

Photo 32. Aerial photograph of mountain front near mile 30.6. Yellow dashed line marks the trace of the main range-front fault; white dashed line marks trace of younger fault that has uplifted the Late Cenozoic fanglomerate (Tf) and may merge with the southern Death Valley fault.

front to the south (Figure 4; Blakely et al., 1999). The Mormon Point Basin reaches its greatest depth of 3 km about 4 km from the mountain front. See p. 14–17 for a more detailed description of the salt pan.

47.0 mi [22.9] For the next several miles, the road traces the distal ends of numerous alluvial fans that spill out of the Black Mountains as shown by Photo 2, page 2. The mountain front displays many features indicative of recent fault movement, including fault scarps in alluvial fans, wineglass canyons, and faceted spurs. These features are described on p. 2–5.

48.8 mi [21.1] Here, the highway nearly runs into the mountain front! Park and walk the short distance to the mountain front to view volcanic rock in the footwall of the frontal fault zone. Although the rock is pervasively brecciated, it displays a prominent layering. Note the many smaller faults that offset this layering.

49.8 mi [20.1] Debris avalanches along mountain front. Two of these debris ava lanche deposits are cut by the frontal fault zone. Knott and Wells (2001) used the approximate age and amount of offset of these deposits to estimate a minimum slip rate of 0.15 to 0.20 mm/year on the fault zone. This rate is far slower than the 3–6 mm/year slip rate estimated on the Northern Death Valley fault zone by Klinger (1999).

50.8 mi [19.1] For the next mile, travellers can see good exposures of shorelines on the steep mountain front, stranded several hundred feet above the valley floor.

53.0 mi [16.9] Excellent view of shorelines rimming the Badwater Spring area. The deposits typically consist of angular pebble to cobble-size rock fragments cemented together with tufa. On the valley-side of the highway, numerous gullies cut the alluvial fan, but run approximately parallel to the fan contours (parallel to the highway and roughly perpendicular to the fan slope). These gullies, visible in Photo 6B, page 10, are liquifaction features from the last major earthquake on the frontal fault zone (Wills, 2001).

53.5 mi [16.4] Badwater Spring, Location G on Figure 14. Parking area and pit toilets on the left. Badwater Spring emanates from the Black Mountains frontal fault zone: looking southward, one can see that it lies on the projected traces of two fault scarps that cut the Badwater fan (Photo 3). The spring discharges between 5 and 10 gallons of water per minute (Hunt, 1975) and supports a thriving, salt-tolerant ecology. The spring hosts a variety of invertebrate fauna, including the endemic Badwater snail (*Assiminea infima*) that forages on algae. It also hosts brine shrimp and brine flies. This location also provides easy access to the salt pan.

The rangefront behind Badwater is structurally complex; there are actually two prominent faults, the range-frontal fault, and the southern part of the Badwater turtleback fault (Photo 33 and see Miller, 1999b). The range-frontal fault likely consists of two discrete surfaces because the range itself trends eastward behind Badwater, but takes an abrupt north turn just beyond. The Badwater turtleback fault lies directly at the edge of the north-trending part of the rangefront, and continues southward through a prominent V-shaped canyon that is visible a short distance up the road. The shoreline deposits depositionally overlie this fault, so one can infer that the Badwater turtleback fault has not been active since they were deposited.

Rocks of the rangefront, below the Badwater turtleback fault, consist of strongly deformed crystalline basement, typical of the Badwater turtleback footwall. Notice that they are cut by numerous light-colored dikes. These dikes consist of ~6Ma quartz latite, which is a rock that roughly resembles rhyolite in composition.

Photo 33. SE-directed aerial view of Badwater area. Continuation of Badwater turtleback fault is shown by the heavy red line within the range; active, frontal fault of Black Mountains is marked by heavy red line at range front.

Directly behind Badwater and above the Badwater turtleback fault, the rocks consist of strongly brecciated Willow Spring Pluton. The Willow Spring Pluton intrudes rocks of the Badwater turtleback deeper within the range. There is also a small fault-bounded block of mylonitic rock above the Badwater Turtleback fault near the edge of the V-shaped canyon (Photo 32).

54.4 mi [15.5] A view up this steep canyon shows that basement gneiss of the Badwater turtleback structurally overlies carbonate rock. Miller (2003) suggested that this older-over-younger relationship was a pre-extensional, basement-involved thrust fault. A view southwards reveals continuation of the Badwater turtleback fault into the V-shaped canyon.

Clear exposures of the Badwater turtleback fault can be seen just north of the highway between here and the next stop.

55.7 mi [14.2] Access to the Badwater turtleback, Location H on Figure 14. Pull-out onto the shoulder and park. Notice that the mountain front steps back from the highway at this spot. Hike 10–15 minutes to the mouth of the southernmost prominent canyon (see arrow on Photo 5A, page 6 for location).

The mouths of each canyon along this part of the Badwater turtleback reveal spectacular exposures of the Badwater turtleback fault and associated fault gouge (Miller, 1996; Cowan et al.,

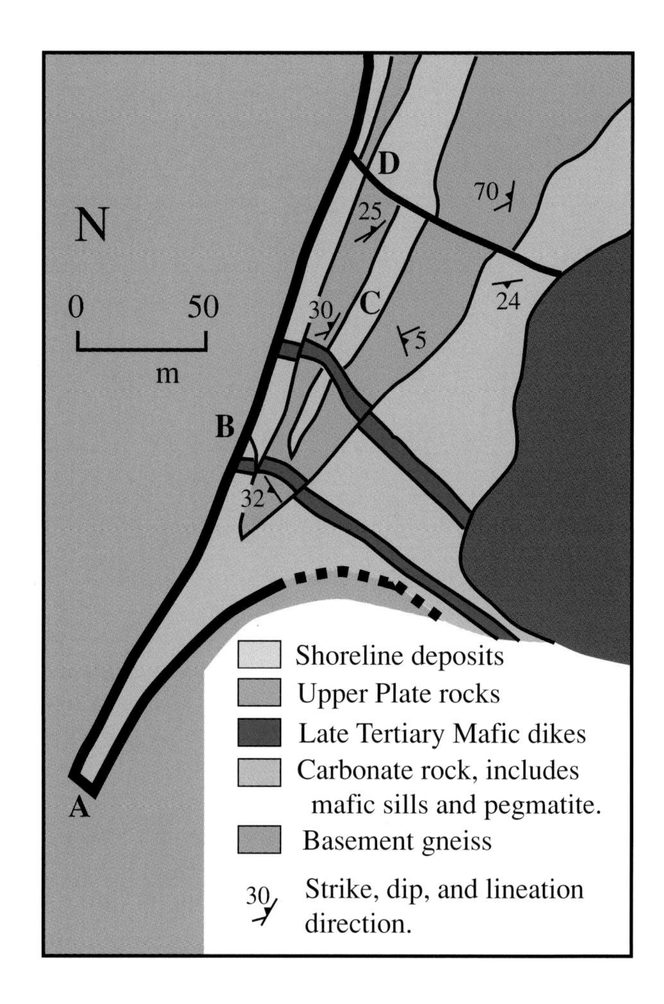

Figure 15. Map of canyon at location H on Badwater turtleback (after Miller in Wright et al., 1991). Capital letters on map indicate points of interest. Location A: exposure of turtleback fault at canyon mouth; B: mafic dikes with chilled margins and cross-cutting relations; C: good exposure of ductile shear zone in marble mylonite. Note that the basement gneiss here structurally overlies the marble as described by Miller (2003). D: High angle fault. The mis-match of unit thicknesses across this fault indicates an oblique component of slip.

2003), as well as some dramatic exposures of mylonitic rocks. It is well worth the effort to take time for exploring.

The first canyon displays ductiley deformed calcite marble with blocks of pegmatite and dolomite, sandwiched between sills of mafic igneous rock. These rocks are locally brecciated and are cut by a series of brittle faults, all of which lie in the footwall, beneath the turtleback fault (Figure 15, location A).

Bypass the dryfall by scrambling up the steeply inclined bench on the south side of the canyon. At the top of this bench, carefully walk into the next part of the canyon to see a wide variety of ductiley deformed footwall rocks, brittle faults, and cross-cutting intrusive relations. Figure 15 is a simplified, and somewhat schematic map of this part of the canyon. Refer to the figure caption for highlights.

57 mi [12.9] Road to Natural Bridge Canyon. This road climbs about 1.5 miles up the alluvial fan to a large parking lot at the mouth of Natural Bridge Canyon. The hike to Natural Bridge itself is only about

1/2 mile, but it is a good idea to continue another 1/4 mile or so, at which point the sandy wash bottom turns to bedrock. Here, the hiker can view the Badwater turtleback fault, which separates the fanglomerate from the metamorphic footwall. The metamorphic rocks consist of calc-silicate marbles (the greenish white rocks) and basement gneiss. The canyon continues for about another 1/8 mile before hitting a 15 to 20 foot high dryfall.

For most of the way to the natural bridge, hikers will walk through fanglomerate that is cut by numerous high-angle faults described in detail by Cichanski (1989) and Nemser (2001). The fanglomerate, which contains a deposit of .77 Ma Bishop ash (Knott et al., 1999) records the unroofing of the Badwater turtleback. Notice that the gravel in the wash bottom and small remnant "terraces" along the edge of the canyon contain both volcanic and metamorphic rock, but the fanglomerate walls contain only volcanic rocks. Therefore, when the fanglomerate was being deposited, the canyon had not yet cut into the metamorphic footwall. However, near the top of the cliffs, one can find an increasing number of metamorphic rocks within the fanglomerate. The presence of these metamorphic rocks indicate the first exposure of the turtleback footwall to erosion.

The natural bridge consists of fanglomerate that extends some 40 feet across the canyon. It appears to occupy the site of an early meander in the canyon, so that erosion through the early canyon wall between each side of the meander loop produced an early "window" for easy passage of water. Continued erosion and downcutting of the canyon enlarged the window to approximately its present shape. Numerous alcoves and slots in the canyon walls attest to the erosional power of flash floods and rainstorms.

59 mi [10.9] Turn-off to Devil's Golf Course. This well-graded dirt road provides an easy access to the salt pan. Devil's Golf Course lies at a slightly higher elevation than the area around Badwater, so it is less frequently flooded (Photo 10B, page 15).

61.4 mi [8.5] Turn-off to Artist Drive. The hill opposite the turn-out contains many fine examples of ventifacts, which are faceted rocks, sandblasted by the wind and sand (Photo 16, page 31).

This 9 mile road provides good views of the multicolored, predominantly volcanic Artist Drive Formation, as well as the Black Mountains frontal fault zone. At a pull-out about 2.5 miles from the entrance, one can see a clear example of the fault surface, complete with striations that indicate its line of slip. As these striations plunge obliquely towards the northwest, they indicate that the latest movement on the fault zone was a combination of right-lateral strike-slip and normal slip. At another pull-out at 3.0

miles, a look westward shows that there is an area of uplift immediately to the west. Here, the visitor is driving through a graben, or a downdropped block between two faults. This graben is visible in Photo 4 (page 4).

The road dips into the washes below two canyons, at miles 2.9 and 3.4. Both of these canyons provide interesting hiking opportunities. As can be seen in photo 4, these canyons are wineglass canyons, indicative of recent uplift of the range front.

At 4.3 miles, a short side road leads to Artist Palette. At this location, a wide variety of colors can be seen in the rock, including greens, purples, and reds. Typically, the reds and purples originate from oxidized iron that is disseminated in the rock whereas the greens originate from reduced iron. The canyon below the view point offers some interesting hiking opportunities.

Beyond Artist Palette, the road descends through gravel deposits in which Knott et al. (1999) found a ~2 Ma tuff to allow its inclusion in the Funeral Formation. At approximately mile 5.5, the road makes a hairpin turn to ascend back towards the mountain front and the underlying Artist Drive Formation. At mile 6.6, the road crosses a fault that separates the Artist Drive Formation from the Funeral Formation.

64.0 mi [5.9] Intersection of West Side Road and highway. A short drive down this road provides easy access to part of the salt pan that is not seen by most visitors.

65.1 mi [4.8] Intersection of Artist Drive exit and highway.

65.5 mi [4.4] Mushroom rock. Although the sign claims that mushroom rock is a ventifact, the actual "sculpting" of the rock took place too far off the ground to have been done by blowing sand. Mushroom rock's shape is more likely a product of weathering brought on by crystallization of salt (Meek and Dorn, 2000).

66.2 mi [3.7] Road to Desolation canyon. A hike up the southeast-trending canyon immediately south of Desolation canyon takes one to the top of the Artist Drive loop. Desolation Canyon is another good example of a wineglass canyon. Note the body of black rock at the dryfall near its mouth. This rock is part of the Ordovician Ely Spring Dolomite that depositionally underlies the overlying Artist Drive Formation.

67.3 mi [2.6] Gower Gulch crosses road. In 1941, water flow down Furnace Creek Wash and was diverted into Gower Gulch at Zabriskie Point. Consequently, Gower Gulch now experiences far more erosion than before. Note how the wash is incised into the fan because of this erosion, and that unlike the other washes along the front of the Black Mountains, this one carries predominantly Paleozoic rocks, all of which originated in Furnace Creek Wash. See mile 3.5, Zabriskie Point.

67.9 mi [2.0] Golden Canyon trailhead. This popular hike leads through a canyon cut into alluvial fan deposits of the Furnace Creek Formation. It continues to Zabriskie Point. Hikers can turn the excursion into a loop by hiking up Golden Canyon to its divide with Gower Gulch, and then following Gower Gulch back to the highway. A trail leads from the bottom of Gower Gulch to the Golden Canyon trailhead.

69.0 mi [0.9] Fault scarps from recent uplift of the Black Mountains.

69.9 mi [0.0] STOP sign at California State Highway 190. End of loop.

Road to Dantes View

This road ends at one of Death Valley's most spectacular view points—Dantes View. On the way, the road traverses a variety of volcanic and sedimentary rocks that range in age from about 14 Ma to 4 Ma. The route is depicted on Figure 14.

0.0 mi Intersection with Highway 190.

0.3 mi Ahead to right, piles of waste rock from Boraxo Pit borate mine.

0.8 mi Straight ahead, the headframe of Billie Borate mine comes into view. The workings and administration building of this mine lie outside the park boundary, but the deposit lies underground within the park. The Billie Mine has been extracting the borate minerals colemanite and probertite since it opened in 1972.

1.4 mi Road to the Boraxo Pit, an open pit borate mine (Photo 3-3, page 19) on the right.

1.8 mi Road bends to right. Above the Billie Mine, the lower part of the Artist Drive Formation is exposed. It contains two tuff beds, dated at 13.7 plus/minus .4 Ma and 12.7 plus/minus .4 Ma (Cemen et al., 1985). The dark gray rocks at the base of the section, visible about 500 feet south of the headframe and just north of a prominent gray pile of dump rock, consists of brecciated limestone of the Bonanza King Formation (Cambrian). Because the contact between it and the Artist Drive Formation is depositional rather than faulted, many geologists argue that rocks of the Furnace Creek Basin formed in approximately the same position as they are today (Wright et al., 1999; Miller and Prave, 2002). See p. 47 for discussion.

2.3 mi Road to Ryan and the Billie Mine lies to the left. From here, the 1920s era mining camp of Ryan is visible; a spectacular angular unconformity is exposed behind it. The unconformity (Photo 28, page 61) shows nearly horizontal basalt flows of the 4 Ma Funeral Formation overlying tilted Furnace Creek Formation (on the south side) faulted against Artist Drive Formation (on the north side).

3.9 mi Road enters canyon cut into basalt of the Furnace Creek Formation. The numerous side canyons expose spectacular cross sections of the lava flows.

5.0 mi On the left are piles of dump rock of light-colored Furnace Creek Formation extracted from underground borate mines.

7.5 mi Trailer Parking and turn-off to the road down Greenwater Valley on left.

9.2 mi Good view to the south of east-dipping volcanic rocks. Uplift along the west-dipping frontal fault zone of the Black Mountains has caused the entire range to tilt eastward. Figure 2 illustrates a schematic cross-sectional view of this tilting.

10.5 mi Road winds through exposures of rhyolite flows and tuffs.

11.3 mi Good exposure of a low-angle normal fault on west side of road. This fault places pink rhyolite flows (above) against a white tuff (below the fault). Inspection of the fault shows a brecciated zone along the fault.

12.5 mi Large pull-out on south side of road with a pit-toilet. Leave trailers here.

13.1 mi Parking lot at Dantes View, Location I on Figure 14, page 66. See p. 12 for a detailed description of Dantes View. Volcanic rocks on either side of Dantes View have been dated by Fleck (1970) as between 6 and 7 Ma.

On the drive down from Dantes View, one gains exceptional views of the Funeral Mountains, Ryan Mesa, and Upper Furnace Creek Wash (Photo 34). The Funeral Mountains consist dominantly of Paleozoic rock that was folded and broken by thrust faults during Mesozoic compression, and tilted and cut by normal faults during Late Tertiary extension. By contrast, Ryan Mesa consists of broadly folded basalt of the Funeral Formation that overlies highly deformed sedimentary rock of the Furnace Creek and Artist Drive Formations. The two areas are separated by the Furnace Creek fault zone, which borders the Funeral Mountains to the southeast (Photo 19, page 40).

Photo 34. View northeastward from near the crest of the Black Mountains across upper Furnace Creek Wash. In the foreground lie east-dipping sedimentary and volcanic rocks, Ryan Mesa occupies the middle ground, and southeast-dipping rocks of the Funeral Mountains lie in the background. The Billie Borate Mine is visible just left-of-center beneath Ryan Mesa; Pyramid Peak, discussed in the Black Mountains road guide, is the high peak on the left. The Furnace Creek fault zone lies between the nearly flat-lying 4 Ma Funeral basalt of Ryan Mesa and the Paleozoic rocks of the Funeral Mountains.

Road to Ubehebe Crater and Scotty's Castle (Northern Death Valley)

From Furnace Creek, Highway 190 skirts the salt pan as it heads north towards Stovepipe Wells. Just past the Salt Creek Hills, the route turns north off of Highway 190 towards northern Death Valley. This road follows the trace of the Furnace Creek-Northern Death Valley fault zone (p. 38–40; Figure 8), along the front of the Grapevine Mountains. As such, it provides excellent opportunities to view offset features from earthquakes along the zone during the last 10,000 years. Klinger (1999) estimated a yearly slip rate of 3 to 6 mm on the fault for the last several thousand years and suggested the most recent earthquake occurred between the years 1640 and 1790. For greater detail of the tectonic geomorphology, the road log by Klinger (2001) is highly recommended.

[Reverse mileage in brackets is from Grapevine Ranger Station]

0.0 mi National Park Service Visitor Center. Turn north on California State Highway 190.

1.3 mi [49.2] Harmony Borax works on left. Turn here for short excursion to Mustard Canyon, a narrow canyon cut into lake deposits of the Furnace Creek Formation. Machette and Slate (2001) mapped the hills around Mustard Canyon to show that they are a Late Quaternary, fault-bounded anticline.

3.0 mi [47.5] Road to Cow Creek. Park offices and employee residence area. Numerous fault scarps cut the alluvial fan just beyond the road. Machette and Crone (2001) estimated the scarps formed between 500 and 840 years ago. Nevares Peak, capped by the Cambrian Bonanza King Formation, forms the prominent peak in this part of the Funeral Mountains. Nevares Spring, one of the more productive springs in Death Valley, issues from the Furnace Creek fault at the base of the peak.

7.0 mi [43.5] Southwest-dipping conglomerate in hills on the east belong to the Funeral Formation. They are faulted against silty lake sediments of the Furnace Creek Formation to the east of here.

10.5 mi [40.0] Beatty cut-off. Turn here to visit the Keane Wonder gold mine, Titus Canyon, or Beatty, Nevada. Refer to page 65 for Beatty cut-off to Titus Canyon road guide.

12.9 mi [37.6] Road to Salt Creek. Salt Creek cuts exposures of the Furnace Creek and Funeral Formations. It is fed by perennial springs near the south edge of the Devil's Cornfield, but dries into a series of marshes during the summer. It hosts the endemic pupfish *Cyprinodon salinus*. See p. 21–22 for a more detailed description of pupfish in Death Valley. Wright and Troxel (1993) mapped the Salt Creek Hills as a northwest-trending anticline. The anticline is cut by the Salt Creek fault on its southwest side.

14.2 mi [36.3] Sea level sign.

17.0 mi [33.5] Intersection of highway with California State Highway 190

17.6 mi [32.9] Intersection with road to Beatty, Nevada. This road, through Mud Canyon, cuts through Quaternary alluvium that overlies Late Tertiary rocks of the Kit Fox Hills.

17.8 mi [32.7] Pull-out on right with pit toilets.

Road continues northward along the west side of the Kit Fox Hills. These hills consist of folded and faulted Tertiary sedimentary rock that trend parallel to the adjacent Furnace Creek-Northern Death Valley Fault Zone. Knott (1999) reported the presence of an ash bed derived from the .655 Ma Lava Creek eruption in Yellowstone in the southeastern part of the hills.

West of the road are small hills that originated as Lake Manly beach bars (Klinger, 2001).

20.1 mi [30.4] Gravel road on left. This road leads to the original Stovepipe Well and the sand dunes.

24.1 mi [26.4] North end of Kit Fox Hills. The road turns right and begins to climb from the valley floor.

25.3 mi [25.2] Mouth of Titanothere Canyon to the right. Bedrock at the front of the Grapevine Mountains here consists mostly of upright Cambrian rocks. In ascending order, they consist of the Zabriskie Quartzite, exposed in the lowest parts of the canyon, the Carrara Formation, Bonanza King Formation, and the Nopah Formation.

28.7 mi [21.8] Pull-out to view an excellent example of a debris flow fan. The best view of the fan is from its north side. There, several different debris flow deposits can be seen, each marked by a slightly different shade of gray or brown, with the darkest shades signifying the oldest deposits. The coloration is a result of desert varnish, a manganese oxide which accumulates on the rocks through time.

32.1 mi [18.4] Titus Canyon Road on the east. This 2.5 mile road climbs the large alluvial fan to meet the one way road down Titus Canyon at a parking lot by the canyon mouth. From there, visitors can hike up the canyon along the one way road, or access Fall Canyon to the north.

From this pull-out, one can see that the thickly bedded Banded unit of the Upper Bonanza King Formation is beneath the thinly bedded Striped unit (Photo 42, page 100). Because the Striped unit member is in fact older than the Banded unit, Reynolds (1966) was able to show that the rocks of this part of the mountain front lie in the overturned, shared limb of the Titus Canyon Anticline and Corksrew Peak Syncline (Figure 16; Photo 26B, p. 59; Photo 39, p. 98). Approximately 3 miles to the southeast, the range front intersects the hinge of the Corksrew Peak Syncline; at that point, the rocks become right-side up. Refer to the Titus Canyon road guide on page 91 for more information about the canyon.

This location also offers a good view of Dry Bone Canyon, directly to the west in the Cottonwood Mountains. Rocks near the canyon mouth consist of the Silurian-Devonian Hidden Valley Dolomite through the Pennsylvanian Bird Spring Formation.

They are intruded by granitic rocks of the Lower Triassic Dry Bone Stock, which are the reddish-brown weathering rocks south of the canyon (Snow et al., 1991).

In general, rocks get older towards the west in the Cottonwood Mountains, an east-tilted fault block. For example, White Top Mountain, which is the high point on the skyline due west of here, is underlain by thickly banded strata of the Ordovician-Silurian Ely Springs Dolomite. Snow (1992), however, mapped numerous pre-Tertiary folds and thrust faults within the range, which cause local departures from this westward age progression. Some of the folding is visible from this location.

Photo 35. View southward over Grapevine Ranger Station and residence area along the Northern Death Valley fault zone. Shutter ridges, developed along the fault can be seen as elongate hills, some of which separate the residence area from the highway.

36.1 mi [14.4] Road crosses alluvial fan from Red Wall Canyon. The mountain front here consists of Tertiary sedimentary rocks that are faulted against Paleozoic limestone of the Grapevine mountains. Reynolds (1966) and more recently Niemi (2002) mapped the Grapevine thrust in this portion of the range, which places Cambrian rock over Ordovician rock.

50.1 mi [0.4] Road to Mesquite Springs Campground on west. The white deposits on the valley floor consist of travertine precipitated from springs along the base of the alluvial fan. The hills to the east and ahead are shutter ridges, uplifted along the Furnace Creek-Northern Death Valley fault zone. They are visible in Photo 35.

50.5 mi [0.0] Grapevine Ranger Station. Telephone, pit toilets, water. The hills behind the residence area lie directly behind the Northern Death Valley Fault Zone.

50.8 mi Road to Ubehebe Crater on left. Turn left. For Scotty's Castle, continue on highway to right for 3 mi. The highway winds through Tertiary volcanic and sedimentary rocks.

51.8 mi Grapevine Ranch Springs come into view on right. The white material is travertine, calcium carbonate deposited from the spring water. These springs issue from small faults and fractures, in lower Paleozoic rock, that lie on trend with the Furnace Creek-Northern Death Valley Fault zone (Steinkampf and Werrell, 2001). Total flow from the dozens of springs and seeps was measured by Miller (1977) at 370 gallons/minute.

Photo 36. Ubehebe Crater as seen from the parking lot. The brightly colored material consists of alluvial deposits, probably of the Miocene Navadu Formation, described by Snow and Lux (1999). The overlying gray material consists of cinders, erupted from the Ubehebe Craters.

53.7 mi Road to Big Pine, California on right. This road takes visitors to the northern reaches of Death Valley, including the Last Chance Range and Eureka Valley. It is accessible by 2 wheel drive vehicles, although the stretch from here to the Last Chance Range is plagued by washboards and small washouts. The road is paved over part of the Last Chance Range and on the west side of the Eureka Valley.

56.5 mi Road to Racetrack Playa on right. High clearance recommended.

56.6 mi Ubehebe Crater. See Young Volcanic Features, beginning on p. 24 and Photo 13A, for a description of Ubehebe Crater.

From here, there are outstanding views south to Tin Mountain, the highest peak of the Cottonwood Mountains, and the Tin Mountain fault, which bounds the west side of the range and cuts directly through Ubehebe Crater. Tin Mountain consists of the Devonian Lost Burro Formation. One can also see the southern part of the Last Chance Range to the west. There, the rocks range from the Cambrian Zabriskie Quartzite at the bottom of Dry

Mountain, to the Devonian Lost Burro Formation, at the top. The area around Dry Mountain was mapped by Burchfiel (1969).

Beatty Cut-off Road to Start of Titus Canyon Road

This road traverses the front of the northern Funeral mountains, with good views of the Chloride Cliff area behind the Keane Wonder mine, and Corkscrew Peak, in the southern Grapevine Mountains. It also crosses a gravel bar, left from Pleistocene Lake Manly.

Beyond Hell's Gate, this road climbs through Boundary Canyon to Daylight Pass and then descends to the Nevada state line and beyond to the Titus Canyon Road. On the way, it passes the structurally complex boundary between the Funeral Mountains to the south and the Grapevine Mountains to the north. This boundary is marked by the Boundary Canyon fault, a detachment fault that has the Grapevine Mountains in its hanging wall, and most of the Funeral Mountains in its footwall.

Most of this route (from about mile 5.0 to 13.0) is depicted on the geologic map by Wright and Troxel (1993) (Map ??).

[Reverse mileage in brackets are from intersection of 190 and Titus Canyon Road]

0.0 mi [23.0] Intersection with Highway 190.

0.5 mi [22.5] View of the Keane Wonder mine area to the northeast; Chloride Cliff lies directly behind and slightly north of the Keane Wonder Mine. White travertine, deposited from springs along the frontal fault zone of this part of the Funeral Mountains occupies the rangefront just north of the Keane Wonder mine (Photo 15A, page 29). The frontal fault is the Keane Wonder Fault, an oblique-slip fault with both right-lateral and normal components of slip.

1.8 mi [21.2] Gravel bar of Lake Manly. This east-west trending feature is about 20' high. It must be younger than the surrounding fan, as its surface is lighter in color, and so has had less staining by desert varnish.

Unlike the material on the alluvial fan, the gravel on this bar is well-sorted and cross-bedded to indicate deposition by water. Well exposed cross-bedding in the gravels is visible in the road cut. Its elevation of 150 feet indicates a lake depth of 430 feet during its time of formation. Hunt and Mabey (1966) noted that the bar extends eastward from an outcrop of Tertiary sedimentary rock. This outcrop was probably an island during formation of the bar.

5.6 mi [17.4] Turn-off to east on gravel road to Keane Wonder mine (Photo 3-1, page 18). This easy road is 2.8 mi long. Park in the parking lot at its end.

Keane Wonder mine. A prospector named John Keane discovered gold in a quartz vein here in April, 1904. By the time the mine closed in August, 1912, it had produced nearly a million dollars in gold. The ore was generally rich and easily worked at the surface but played out at depth. This mine hosted an aerial tramway, much of which is still intact, that connected the mine workings up high, with the stamp mill at the mountain front. According to Lingenfelter (1986) the tramway itself became an attraction. It operated by gravity: energy from descending ore buckets, which weighed about 1/2 ton when full, would pull the empty buckets up to the top. The tramway earned the name "sky railroad" as visitors would occasionally ride it up to the mine.

The Keane Wonder Mine area hosts some of the most interesting metamorphic rocks in the Death Valley region. They consist of marble, high grade pelitic schist, and amphibolites. The schist locally contains kyanite and staurolite and with the marble originated as the Proterozoic Crystal Spring Formation; the amphibolite originated as diabase sills (Wright and Troxel, 1993). Both of these units are described in more detail on p. 51, 53. To the north, at Monarch Canyon, the schist contains sillimanite to indicate even greater temperatures of metamorphism. Metamorphism likely occurred during the Late Cretaceous at depths of about 20 km, based on pressure estimates of 6.5 +/– 1 kbar and and temperatures of 450–700°C (Labotka and Albee, 1988). The rocks experienced significant retrograde metamorphism, probably during the late Tertiary extension, as shown by extensive alteration of mafic minerals to chlorite.

5.8 mi [17.2] Good exposures of south-dipping Miocene sandstone to the west. As the road continues northward, more exposures come into view on the east.

8.0 mi [15.0] Ahead and slightly west of the highway, are the peaks of Death Valley Buttes, which consist of sandstone and siltstone of the Cambrian Wood Canyon Formation.

8.5 mi [14.5] View of Monarch Canyon to the east. Monarch Canyon (see description at mile 13.3) is cut into a highly deformed section of Crystal Spring Formation.

9.3 mi [13.7] One can see a good view of the Keane Wonder fault looking back to the southeast. There, the Keane Wonder fault separates the Proterozoic and Paleozoic rock of the Funeral Mountains from downdropped Tertiary sedimentary rock.

9.9 mi [13.1] Join main highway to Beatty at Hell's Gate. Large pull-out and pit toilet. Good view of the recumbent syncline at Corkscrew Peak. Rocks on east side of highway consist of the Proterozoic Stirling Quartzite.

Turn right. (northeast)

10.8 mi [12.2] From here until the next stop, one can view the Boundary Canyon fault on both sides of the road as the contact between gray-green rocks below and tan rocks above. The green rocks are Late Proterozoic Johnnie Formation and the tan rocks are part of the Late Proterozoic Stirling and Wood Canyon Formations (see discussion on pp. 35–36 and Photo 17). Structurally beneath this fault is another detachment, the Chloride Cliff fault (Wright and Troxel, 1993). The Chloride Cliff fault separates the overlying Johnnie Formation from underlying Crystal Spring Formation.

12.0 mi [11.0] Pull-out to inspect Boundary Canyon fault.

13.3 mi [9.7] Gravel Road on south that leads to Monarch Canyon and Chloride Cliff. The road is 2 WD accessible (high clearance recommended) as far as Monarch Canyon (2.2 mi), but 4 WD after that. Besides Monarch Canyon, the 2WD part of the road accesses exposures of the Titus Canyon Formation.

4 WD

Monarch Canyon. Monarch Canyon contains outstanding exposures of ductiley deformed marble, schist, and gneiss in the footwall of the Boundary Canyon and Chloride Cliff faults. Many of the ductile features formed during uplift of the metamorphic rocks from mid-crustal depths along the Boundary Canyon fault. To visit, either park at the hairpin turn at 2.2 mi and walk into the canyon, or drive the additional .5 mi into the canyon and park above the first dryfall. The hairpin turn is right at the Boundary Canyon fault, between Stirling Quartzite in the hanging wall and Johnnie Formation in the footwall. The fault here, however, is poorly expressed. Approximately 1/4 mile into the canyon, the Chloride Cliff fault separates Johnnie Formation in the hanging wall from Crystal Spring Formation in the footwall.

At the dryfall, proceed on the trail down canyon to more exposures of ductiley deformed Crystal Spring Formation and eventually basement gneiss. The gneiss is one of only two known exposures of basement rock in either the Funeral or Grapevine mountains. Basement rock of Death Valley is described in more detail on p. 50–51. Many of the schistose rocks in the Crystal Spring Formation contain sillimanite, as described by Labotka and Albee, (1988). Other more quartzitic rocks are strongly mylonitic. The canyon continues for about a mile before cliffing out at an impassable waterfall.

The side canyon that joins Monarch Canyon at the first dryfall contains numerous small-scale folds in carbonate rock of the Crystal Spring Formation.

15.1 mi [7.9] The two gray peaks on the left side of the road consist of Cambrian Bonanza King Formation. According to Troxel (1974), these peaks are actually gigantic clasts of the Bonanza King Formation within landslide deposits of the Titus Canyon Formation.

16.1 mi [6.9] Daylight Pass. The craggy peaks on both sides of the road are the Cambrian Zabriskie Quartzite overlying the Wood Canyon Formation. Overlying these rocks is the Oligocene Titus Canyon Formation, which occupies the area of lowrelief to the north. Overlying the Titus Canyon Formation are 22–20 Ma felsic volcanic rocks.

16.6 mi [6.4] Northeast-tilted volcanic rocks straight ahead.

18.0 mi [5.0] Entering the upper reaches of the Amargosa Valley. There is a good view of Bare Mountain, a highly faulted block of Lower Paleozoic rock, straight ahead. The Bullfrog Hills are to the northeast. Slightly farther east sits the ghost town of Rhyolite.

20.4 mi [2.6] California-Nevada State Line.

23.0 mi [0.0] Titus Canyon Road on north side of highway.

Titus Canyon

This one-way 2WD (high clearance recommended) road takes visitors across the Grapevine Mountains, beginning in the upper reaches of the Amargosa Valley near Rhyolite, and ending in Death Valley. Both the geology and scenery are spectacular. The trip begins in Upper Tertiary volcanic rocks, descends through the Oligocene Titus Canyon Formation, and continues down through Paleozoic strata. The Paleozoic strata are deformed into a mountain-scale overturned fold, so the lower reaches of the canyon are within overturned rock. The general geology is shown on Figure 16.

0.0 mi Intersection of Nevada State Highway 374, approximately 6 miles west of Beatty.

For the next several miles, the road approaches the Grapevine Mountains. In general, the nearby light and dark multicolored peaks and hills are underlain by Tertiary volcanic rocks, whereas the craggy peaks behind them consist of Paleozoic sedimentary rocks. Of these, the red-colored peaks are part of the Cambrian Zabriskie Quartzite (p. 54–55).

The volcanic rocks consist predominantly of felsic lava flows and welded tuffs that are about 22 to 20 million years old (Reynolds, 1976). The source for many of these rocks was at or

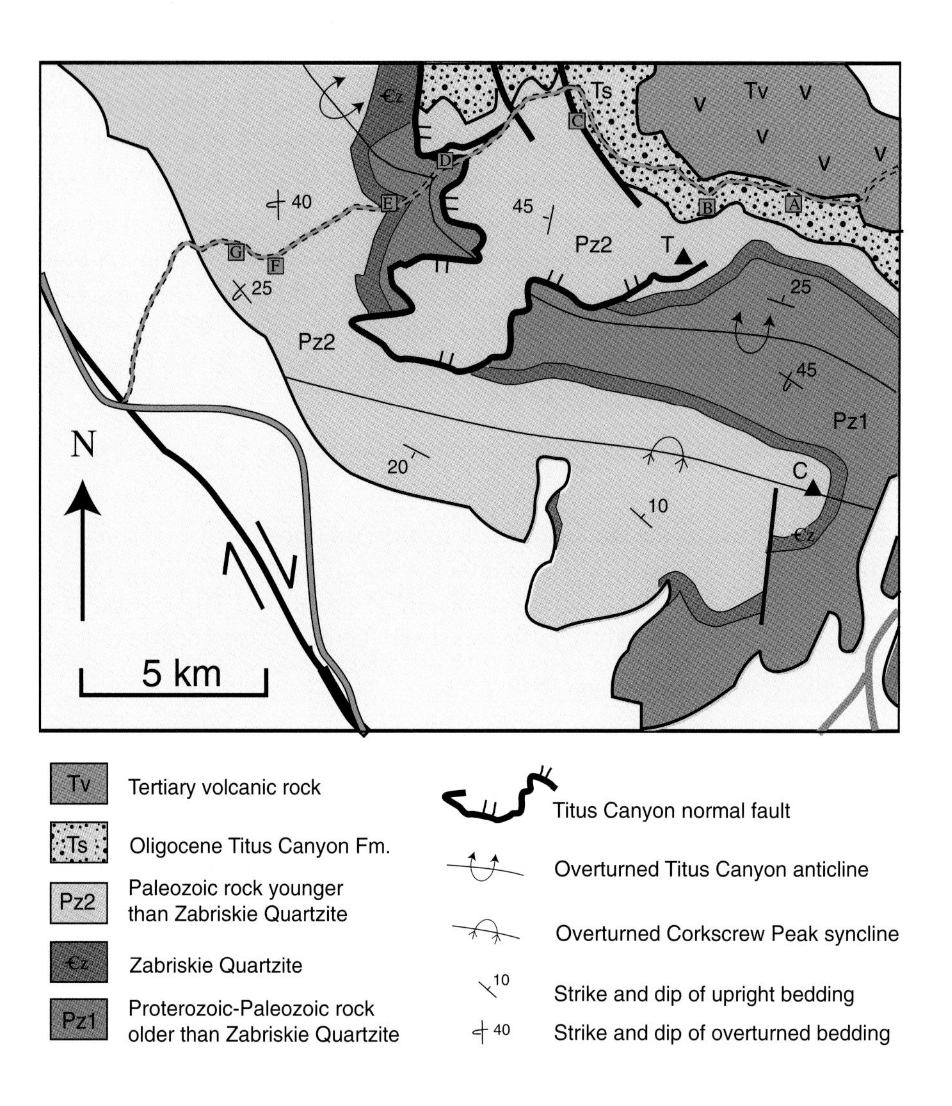

Figure 16. Simplified map of Titus Canyon area, based on detailed mapping by Reynolds (1969). Recommended stops shown by letters in green squares. Abbreviations consist of C: Corkscrew Peak; T: Thimble Peak.

near the Nevada Test Site to the east. Many of the flows are classified as latite because they contain visible crystals of both potassium feldspar and plagioclase, but little or no quartz. The latite flows possibly originated from a latite plug mapped by Reynolds (1969) due north of Red Pass (mile 12.4).

At about 6.0 mi, the road enters hills underlain by Tertiary volcanic rock.

7.5 mi — Gully on right contains an outcrop of the Tertiary volcanic rocks. A white tuff with numerous rock fragments is overlain by a younger deposit with cross-beds. The cross-beds indicate that this younger deposit was re-deposited by running water.

From here to White Pass, Paleozoic rocks (mostly the reddish Zabriskie Quartzite) lie to the west, while Tertiary volcanic rocks lie to the east.

Photo 37. View westward from White Pass.

9.6 mi White Pass—Location A on Figure 16. Pull-out and view the upper reaches of Titanothere Canyon. Here, one can gain a perspective on the stratigraphy of the field trip: the peaks along the east side are Tertiary volcanic rocks, the peaks on the west side are Lower Paleozoic rocks, described in more detail on p. 54–56, and the subdued topography in between is underlain by the Oligocene Titus Canyon Formation. Photo 37.

The Titus Canyon Formation consists mainly of green-colored conglomerate, sandstone, and red and green shale, deposited in river and lake environments. It also contains limestone megabreccia deposits that probably originated as landslides. Because landslides require steep topography, their presence provides indirect evidence for a period of mountain-building during the Oligocene. In the 1930's, Donald Curry, the first park naturalist, discovered the skull of a Titanothere in the Titus Canyon Formation. A cast of the skull is on display in the Death Valley Visitor Center. More description of the Titus Canyon Formation and its environment are on p. 60.

Stop here and turn around if it's raining. The road below traverses shale of the Titus Canyon Formation, which becomes dangerously slick when wet.

10.3 mi Pull-out. From here, road descends a steep grade through interbedded shale and conglomerate of the Titus Canyon Formation.

11.1 mi Road makes a hairpin turn to the north. Good view up gully to a dark-colored blocky peak. This peak is a volcanic plug, left behind after erosion stripped away the surrounding material. A short hike up the gully will reveal that the rock contains flow-banding that is vertical, to indicate flow of magma upwards through the volcanic conduit. Another good access for the plug is on the other side of the pass at mile 13.1. Photo 38.

Photo 38. Aerial view of volcanic plug in vicinity of Red Pass. Plug is the dark, olive-green block immediately left-of-center.

11.6 mi Road crosses a fault and enters rock consisting of limestone megabreccia. These are some of the deposits interpreted by Reynolds (1974) as having originated in landslides.

12.4 mi Red Pass, Location B on Figure 16, Photo 27. The red rock here belongs to the basal part of the Titus Canyon Formation. It consists mostly of red sandstone and conglomerate. Outcrops of the limestone megabreccia exist a short way down the road (~100 yards).

There is a small pull-out here and an awesome view into the upper drainage of Titus Canyon. The geology resembles that below White Pass in that Upper Tertiary volcanic rock dominates the east, Paleozoic rock (Cambrian through Ordovician) dominates the west, and Tertiary Titus Canyon occupies the valley in the middle.

Most of the Paleozoic rock consists of dark gray limestone of the Bonanza King Formation. Near the contact with the overlying Titus Canyon Formation, one can see that the Bonanza King Formation is folded into an anticline. However, neither the Titus

Canyon Formation, nor the Upper Tertiary Volcanic rocks are folded at that locality. There is a fault between the folded Paleozoic and non-folded Tertiary rocks.

The hike to Thimble Peak, one of the high points of the Grapevines, is less than 2 miles away and a 1500 foot elevation gain from Red Pass. From there, hikers get an expansive view of Death Valley, which on clear days includes both the Sierra Nevada and Badwater. If attempting this hike, parking in the pull-out below is recommended, as parking at Red Pass is both popular and limited.

12.45 mi Outcrops of limestone megabreccia on left. No parking. View of latite plug to the northwest.

12.8 mi Pull-out with room for one vehicle.

13.1 mi Road makes sharp turn. Park here (space will fit two small vehicles) to access latite plug. Walk up gully and scramble up steep slope.

15.5 mi Leadfield, location C on Figure 16. Ample Parking and pit toilets.

Leadfield marks the site of a large, but low grade body of lead ore that was promoted on the Los Angeles stock exchange in January, 1927 as a high grade body, accompanied by exorbitant and sometimes fraudulent claims of its value (Lingenfelter, 1986). The promoters even constructed the initial road up Titus Canyon to showcase the area to potential investors on a one day tour. Of course the tour was a success; investors, who lacked the expertise to assess the true value of the deposit had to rely on the word and apparent confidence of the promoters. No lead was actually produced from the mine, but many investors lost a great deal of money.

Behind the buildings, one can see the Bonanza King Formation dips steeply eastward. It is faulted against the less steeply dipping Titus Canyon Formation. To the northeast, volcanic rocks display numerous small normal faults.

15.9 mi Road turns left and crosses fault into Bonanza King Formation. Here is an outstanding example of apparent dip: the rocks appear to be folded into a syncline, but in fact, they are all dipping in approximately the same direction. This illusion occurs because the canyon makes a sharp bend at this location, so the canyon walls intersect the bedding at two different angles.

16.2 mi Entering Titus Canyon itself. Until this point, the road has been in a tributary of Titus Canyon.

17.2 mi Numerous small normal faults can be seen displacing beds of the Carrara Formation on the right.

18.0 mi Petroglyphs and Klare Spring; location D on Figure 16. Good pull-out on left. The petroglyphs can be found on the down-road side of the large limestone boulder at road level, just below the interpretive sign. Notice that the boulder is cemented in place by travertine, a typical deposit of spring water. The site of the present spring lies within the riparian vegetation, a short distance to the west. This spring produces about 22 gallons of water/minute (Steinkampf and Werrell, 2001). As the biggest source for water in the area, it plays a central role in the local ecology.

Klare Spring issues from the Titus Canyon fault, a low-angle normal fault that dominates much of the structural geology of Titus Canyon (Fig. 16). In general, the fault places the Cambrian Bonanza King Formation above the fault against older rocks below the fault. In upwards succession, these older rocks consist of the Wood Canyon Formation, the Zabriskie Quartzite, the Carrara Formation, and the Lower Bonanza King Formation. Here, the Titus Canyon fault is visible as the contact between the gray limestone of the Bonanza King Formation in the hanging wall, and the brownish sandstone of the older Late Proterozoic-Cambrian Wood Canyon Formation in the footwall. An orange stain, from travertine deposition, marks the fault in places on the other side of the canyon. Strongly deformed rocks of the Wood Canyon Formation exist right along the road, immediately below Klare Spring. The rugged terrain in this area was mapped in detail by Reynolds (1969).

18.4 mi View to the south of the Titus Canyon fault.

18.9 mi Alluvial terrace. This deposit demonstrates the sometimes complicated cycling of deposition and erosion in canyons. The top of this deposit marks the canyon bottom at an earlier time, but since then, the canyon has cut down to its present elevation. However, for the alluvial materials to be deposited in the first place, the canyon already needed to be at least as deep as it presently is.

Deposition and erosion in Titus Canyon, and in most other canyons on the east side of Death Valley, is controlled by the faulting at the front of the range. With each earthquake, the range rises somewhat and increases its gradient at the canyon mouth. Simultaneously, the range tilts slightly eastward, which decreases the gradient behind the canyon mouth. This decrease in gradient causes deposition in the higher reaches of the canyon. However, the increased gradient at the canyon mouth gradually works its way up canyon through erosion to remove the deposited materials and cut the canyon deeper.

The red-colored peaks on the north skyline are Cambrian Zabriskie Quartzite.

19.4 mi Pull-out to view Zabriskie Quartzite and Titus Canyon anticline; location E on Figure 16. The pink-red Zabriskie Quartzite dips steeply eastward where it crosses the road at this point. However, it exhibits abundant cross-bedding that indicates it gets younger to the west, which means that it is overturned. The stratigraphically underlying Wood Canyon Formation, which lies just up the road from here, is resting on top of these rocks.

By following the Zabriskie Quartzite and Wood Canyon Formation upwards to the ridge line north of the road, one can see that they bend back into an upright position. This structure is the Titus Canyon Anticline, shown in Photo 39 and depicted on Figures 16 and 17. The anticline controls the structural geology of the trip down the rest of the canyon, as all the rocks lie in its overturned limb.

The prominent notch near the east end of the ridge marks the Titus Canyon fault. There, the Zabriskie Quartzite is faulted against the Bonanza King Formation.

19.6 mi Overturned Carrara Formation forms the canyon on the left.

19.8 mi Road crosses contact into overturned Bonanza King Formation.

20.1 mi Antiformal syncline and synformal anticline in canyon wall on left. As the rocks here are in the overturned limb of the much larger Titus Canyon Anticline, these folds are upside down. Therefore, the antiform is actually a syncline and the synform is an anticline.

21.9 mi Road enters narrows of Titus Canyon. Note the increasingly complex folding in the Bonanza King Formation.

22.2 mi Good pull-out and view of tight folding in the Bonanza King Formation; location F on Figure 16. At first glance, the rocks in the short stretch of canyon below here appear to be nearly horizontal and so relatively undeformed. However, on closer inspection, one can see that they are tightly folded, and that the main folds can

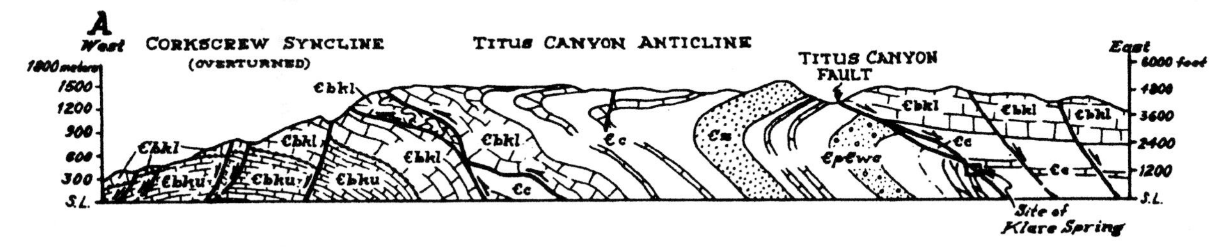

Figure 17. NE-SW cross-section of Titus Canyon area approximately parallel to road of Fig. 16 beginning about 3 km upstream from Klare Spring (from Reynolds, 1976). Note that Klare Spring issues from Titus Canyon fault, and that lower part of canyon lies in the overturned limb of the Titus Canyon anticline.

Photo 39. Titus Canyon Anticline. Here, rocks of the Wood Canyon Formation, Zabriskie Quartzite ("z" in photograph), Carrara Formation, and Bonanza King Formation become overturned along the ridge crest in the middle ground; along the Titus Canyon road near the bottom of the photo, one goes topographically downward into younger and younger rock. Because of this fold, many of the seemingly flat-lying rocks near the mouth of Titus Canyon are actually upside-down!

be followed for some distance back up canyon. The complicated geometry is a product of the irregular fold shapes. Rather than having straight hingelines, these folds appear to have strongly curved hinges.

23.1 mi Brecciated rock and evidence for flood erosion; location G on Figure 16. The wall at this spot is a mosaic of broken limestone blocks re-cemented with white calcite. Many of the blocks do not touch each other, but appear to be floating in the white calcite. It is therefore likely that the fracturing and calcite mineralization occurred in many stages, each of which gradually separated individual blocks. The breccia is probably not a product of faulting, however, because there are no obvious faults here. Instead, it may be related to progressive fracturing and calcite mineralization during late stages of folding. More zones of breccia exist farther down canyon.

Photo 40. Canyon narrows at mile 23.3. Note smooth, scoured appearance of canyon walls. Numerous small normal faults offset beds in the background.

From here to the canyon mouth the evidence for flood erosion is spectacular. Look for striations on the canyon walls (the breccia clearly displays striations) that formed by the abrasive action of rocks and boulders moving down canyon. Many of the walls are scalloped from the plucking of loose material from the wall during floods, and also have a fluted appearance from abrasion. Photo 40 shows this scalloped effect at mile 23.3. Just below the breccia, there is a large alcove cut into the canyon wall.

23.9 mi Mouth of canyon, parking lot, and pit toilet. Begin two-way road. A hiking trail from the parking lot leads to Fall Canyon, the next canyon north.

26.5 mi Intersection of Titus Canyon road with main highway. Just before the highway, the Titus Canyon road climbs over a steep berm, which is a fault scarp on the Northern Death Valley fault zone.

Photo 41. Narrows of Titus Canyon.

Photo 4 2. Front of Grapevine Mountains. Narrow mouth of Titus Canyon lies near center of photo at end of road; wider mouth of Fall Canyon lies left (north) of Titus Canyon. Striped member of Bonanza King Formation is older than Banded member.

From here, it is instructive to view the rangefront and see stratigraphic evidence that the rocks are in fact overturned (Photo 42). The mouth of Titus Canyon is in the Striped Unit of the upper member of the Bonanza King Formation, which from a distance is distinguished by narrow banding; 546 yards to the south, one can see that it overlies the Banded Unit, which from a distance is distinguished by thick banding (Reynolds, 1969). However, the Banded Unit is actually younger than the Striped Unit.

Western Park Boundary to Stovepipe Wells (Western Death Valley)

This road traverses many features that reflect the crustal extension that formed today's landscape. These features include Pliocene basalt, Late Tertiary sediments of the extensional Nova Basin (p. 42), and several large normal faults, including the range-front fault of the Panamint Mountains. At Father Crowley overlook, there is a spectacular glimpse of Death Valley's older history.

[Reverse mileage in brackets is from Father Crowley Overlook]

0.0 mi [4.7] Park boundary on Highway 190.

0.5 mi [4.2] View of cinder cones of the Darwin Plateau and the flat-topped Hunter Mountain ahead to the left. Hunter Mountain, to the east, is underlain by Middle Jurassic granitic rocks of the Hunter Mountain Batholith (Chen and Moore, 1982). Panamint Butte, at 1 o'clock, is capped by Pliocene basalt, similar in age and character as the basalt here (Burchfiel et al., 1987).

1.0 mi [3.7] View of Telescope Peak, ahead and to the right. Roadcuts here expose colluvium (gravity-driven erosional materials) and basalt.

2.1 mi [2.6] To the east is a deep canyon that cuts into gently tilted basalt flows. Paleozoic rocks beneath the basalt are more steeply tilted to indicate deformation before eruption of the basalt.

3.8 mi [0.9] Surrounding landscape covered by red basaltic cinders. The red color results from oxidation of iron in the cinders. In contrast, basaltic lava flows retain their black color for much longer periods of time, as they are far less accessible to weathering processes than cinders.

4.7 mi [0.0] Turn-out to Father Crowley overlook. Memorial plaque, placed by the Death Valley 49er's (an organization of Death Valley supporters), describes Father John Crowley (1891–1940) as the "Padre of the Desert." Drive down the gravel road .5 mile to a spectacular view of Panamint Valley and the western Panamint Range.

Father Crowley overlook. The Panamint Valley resembles Death Valley in that it is down-dropped along a system of oblique normal faults that borders the east side of the valley. Numerous fault scarps in the alluvium record recent activity on the fault. At the north end of Panamint Valley, the fault bends to the northwest to become the right-lateral Hunter Mountain Fault. Burchfiel and others (1987) estimated 5 to 6 miles of right-lateral slip on this fault.

Looking northeastward and eastward, the Panamint dunes at the north edge of Panamint Valley, the fault-controlled rangefront behind the dunes, and Panamint Butte can be seen. The rangefront directly behind the dunes consists of the middle Jurassic Hunter Mountain Batholith. By contrast, Panamint Butte is underlain by the highly striped rocks typical of the Pennsylvanian and Permian rocks, and is capped by Pliocene basalt, similar in age to the basalt at this overlook. The intrusive contact between batholith and the Paleozoic rocks is marked by the color change from the greenish intrusive rocks to the tan and gray colors of the sedimentary rocks. At Panamint Butte, the Lemoigne Thrust places rocks as old as Cambrian over the Pennsylvanian-Permian Keeler Canyon Formation. An overturned syncline can be seen in the Keeler Canyon Formation beneath the thrust just more than halfway up the mountain. Panamint Butte was mapped by Hall (1971).

South of Panamint Butte is the rounded crest of Tucki Mountain. Still farther south, Telescope Peak makes up the high point in the Panamint Range. The Panamint Range has been uplifted along the Panamint fault zone and tilted eastward. On the west side of Panamint Valley there are east-dipping Paleozoic rocks of the Argus Range.

Just below the overlook, the view down Rainbow Canyon shows some geologic relations that encapsulate much of the geologic history of the Death Valley region. A better view (Photo 42) can be gained by taking the short hike to the ridge from mile 2.6 below the overlook. There, Pliocene basalt overlies Mesozoic granitic rock that intrudes folded Paleozoic limestone. The granitic rock underlies the gray-green steep bouldery slopes directly across the canyon. A basaltic dike cuts through the Paleozoic rock and appears to merge with the lava flows. From this view, the following sequence of events can be inferred: deposition of limestone in a marine environment during the Paleozoic Era (p. 56); folding from crustal compression during the latest Paleozoic or Mesozoic (p. 57); intrusion of the first plutons related to the Cordilleran magmatic arc of the Mesozoic (p. 58); volcanic activity and faulting related to crustal extension in the Late Tertiary (p. 60).

Return to highway and *reset odometer*. Turn left (east).

Reset Odometer 0.0

[Mileage from Stovepipe Hills, 38.2]

0.3 mi [37.9] View of Father Crowley overlook. The overlook consists of a thin layer of basalt over Paleozoic limestone. Numerous mafic dikes, likely part of the Jurassic Independence swarm, cut through the limestone.

1.3 mi [36.9] Pull-out at outcrop of Paleozoic limestone beneath basalt. Note the altered mafic dike.

2.5 mi [35.7] Large pull-out on right (south) opposite a road cut through a cinder cone. The numerous fractures in the cinder cone are filled with white calcite.

2.6 mi [35.6]

Large pull-out on right. Walk to east end of road cut and up to ridge for an outstanding view of Rainbow Canyon (Photo 43). From here, evidence for the principal geologic events that shaped the Death Valley region can be seen. See description for Father Crowley Overlook, page 100.

4.9 mi [33.3] Road goes through gash in volcanic cinders and flow breccias.

6.1 mi [32.1] Pull-out and road cut. Here, alluvial fan deposits stratigraphically underlie basaltic lava flows to indicate that mountain-building had been taking place prior to eruption of the basalt. 55 yards or so down the road, basaltic feeder dikes cut through the alluvial fan deposits. Darwin Canyon lies just south of the road.

6.7 mi [31.5] Exposure of granitic rock. Here, the same type of rock that intrudes the folded limestone of Photo 43 can be closely inspected.

6.8 mi [31.4] Road to Darwin Falls.

Photo 43. View northeastward of rocks in Rainbow Canyon. Folded Paleozoic rock is intruded on left side by Mesozoic granitic rock of Hunter Mountain Batholith. These rocks are overlain unconformably by Pliocene basalt. Normal faults cut the entire section. Note the basaltic dike near the eastern edge. In the background, rocks of the Hunter Mountain Batholith (HMB) intrude Paleozoic rocks (Pz) near Panamint Butte.

7.8 mi [30.4]	Panamint Springs Resort. The resort offers good food and camping. From here, the road descends the bajada and crosses Panamint Valley.
10.3 mi [27.9]	Intersection with road to Trona.
12.3 mi [25.9]	Intersection with gravel road to northern Panamint Valley. The low hills immediately west of this road are underlain by Paleozoic limestone. Many limestone boulders along the base of the hills have been sand-blasted into spectacular ventifacts (p. 31).
14.5 mi [23.7]	Cross frontal fault zone of Panamint Mountains. The road steepens considerably as it leads up into the Panamint Mountains. Enroute, observe the Late Tertiary basalt, as well as uplifted and locally faulted alluvial fan deposits and lake beds of the Nova Formation in roadcuts. The Nova Formation was deposited in the extensional Nova basin prior to development of modern Death Valley (Photo 21B on p. 44).
17.4 mi [20.8]	Pull-out with radiator water. From here, observe tilted Nova Formation straight ahead and the Sierra Nevada Range to the west.
19.3 mi [18.9]	Normal faults cut interbedded basalt and alluvial fan deposits. There is a large pull-out just past the roadcut for those who want to inspect the faults.
21.5 mi [16.7]	Towne Pass, elevation 4956 feet. Large pull-out and good view of northern Tucki Mountain. Here one can see the Towne Pass fault, which dips moderately westward beneath predominantly alluvial fan deposits of the Nova Formation. A **megabreccia**, consisting of mostly of Paleozoic carbonate rock that was broken and re-deposited as part of the Nova Formation (Wernicke et al., 1989), lies in the footwall.
26.4 mi [11.8]	Tertiary basalt crops out on both sides of the highway. Beyond it to the south are exposures of Nova Formation.
29.0 mi [9.2]	Road to Wildrose Canyon.
29.2 mi [9.0]	Emigrant Ranger Station.
31.0 mi [7.2]	Tertiary basalt flows to south are interbedded with sedimentary rocks of the Nova Formation.
32.8 mi [5.4]	Black Point
34.5 mi [3.7]	View to the south shows the Nova Formation faulted against the Proterozoic Kingston Peak Formation. This fault is the Emigrant normal fault, described in detail by Wernicke et al., 1986.

Megabreccia. A very coarse, large-scale breccia.

36.5 mi [1.7] Good View of Mosaic Canyon normal fault to the southeast. It separates Proterozoic Johnnie Formation in its hanging wall from Proterozoic Noonday Dolomite in its footwall.

38.0 mi [0.2] Road to Mosaic Canyon.

Mosaic canyon provides outstanding examples of canyon erosion and deposition as well as processes of structural geology. The canyon narrows, which are only a few minutes walk from the parking lot, are carved into marble of the Proterozoic Noonday Dolomite and recent deposits of several overlapping mudflows. Where marble is exposed, scouring from flash floods in the canyon gives the walls a smooth polished appearance; where mudflow deposits are exposed, scouring of the many interlocking clasts have created stunning natural mosaics (Photo 44).

The narrows lie immediately beneath the Mosaic Canyon fault, a low-angle normal fault that juxtaposes the Noonday Dolomite with the overlying Johnnie Formation. The fault is exposed on the east side of the parking lot, where it contains a thick zone of cataclasis and fault gouge. In the canyon narrows, however, the Noonday Dolomite contains both brittle and ductile features, some of which relate to slip on the fault, and some of which formed earlier (Hodges et al., 1987). These structures include a well-developed foliation, suggestive of high strains in a ductile shear zone, isoclinal folds, a second generation of re-folded folds, small-scale brittle-ductile faults, boudins, and brittle normal faults.

Photo 44. Canyon narrows in Mosaic Canyon. In this photo, the left wall and canyon floor consist of polished Noonday Dolomite, whereas the right, overhanging wall consists of a polished mudflow deposit.

38.2 mi [0.0] Stovepipe Wells. Gas, general store and toilet facilities, motel accommodations and restaurant.

40.4 mi Good access to Mesquite Flat sand dunes (Photo 14A) some of Death Valley's most dramatic, and most accessible sand dunes. See pp. 25–28 for a discussion of sand dunes in Death Valley.

44.1 mi Devil's Cornfield. A high groundwater table in this area has created a meadow-like quality here. The dominant plant, arrowweed, grows on mounds of sand.

45.4 mi Intersection with road to Ubehebe Crater and northern Death Valley. If turning left, refer to road guide for northern Death Valley on page 83.

Map X. Bedrock geologic map of the Racetrack Playa area, from McAllister, 1956. North is to the top; map area measures approximately six miles from north to south. Abbreviations of the most prevalent rock units are as follows: Cr: Racetrack Dolomite (Cambrian); Cn: Nopah Formation (Cambrian); Op: Pogonip limestone (Ordovician); Oe: Eureka Quartzite (Ordovician); Oes: Ely Springs Dolomite (Ordovician); DShv: Hidden Valley Dolomite (Silurian-Devonian); Dlb: Lost Burro Formation (Devonian); Mtm: Tin Mountain Limestone (Mississippian); Mp: Perdido Formation (Mississippian); PAL: Undifferentiated Paleozoic Limestone; Kh: Hunter Mountain quartz monzonite; Khp: Hunter Mountain quartz monzonite with large orthoclase crystals.

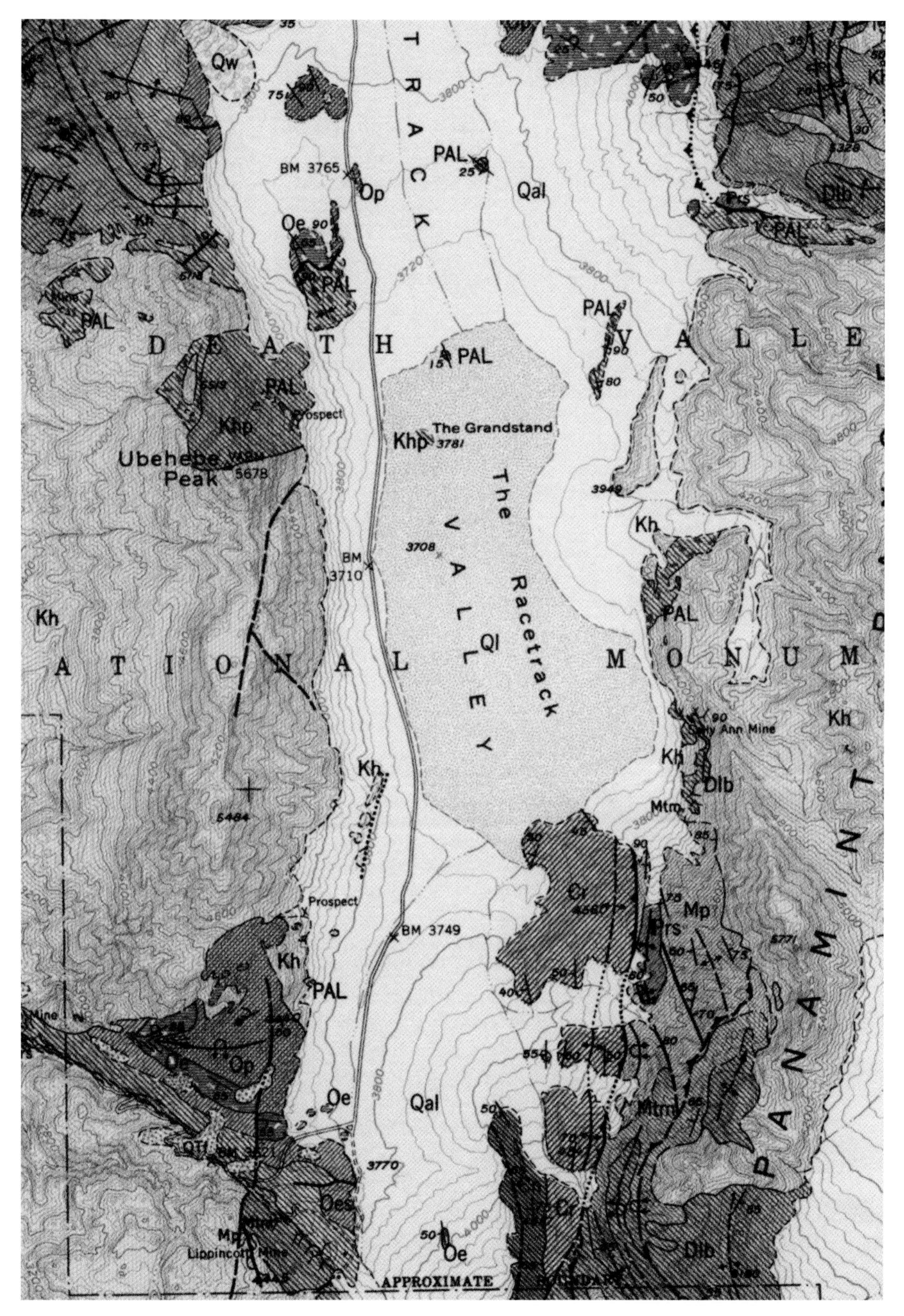

Map Y. Bedrock geologic map of the area near Jubilee Pass, from Wright and Troxel, 1984. North is to the right; map is approximately 2.5 miles from north to south. Location "B" is Exclamation Rock; location "C" is the exposure of the Amargosa fault in Virgin Spring Canyon. Abbreviations of the most prevalent rock units are as follows: pCg (brown): Precambrian basement rock; db (purple): Precambrian diabase dikes; pCcl, pCcu: Lower and Upper Crystal Spring Formation respectively; pCnl (light blue) and pCnu (green): Lower and Upper Noonday Dolomite respectively; Qtfc (yellow): Funeral Formation (Tertiary-Quaternary).

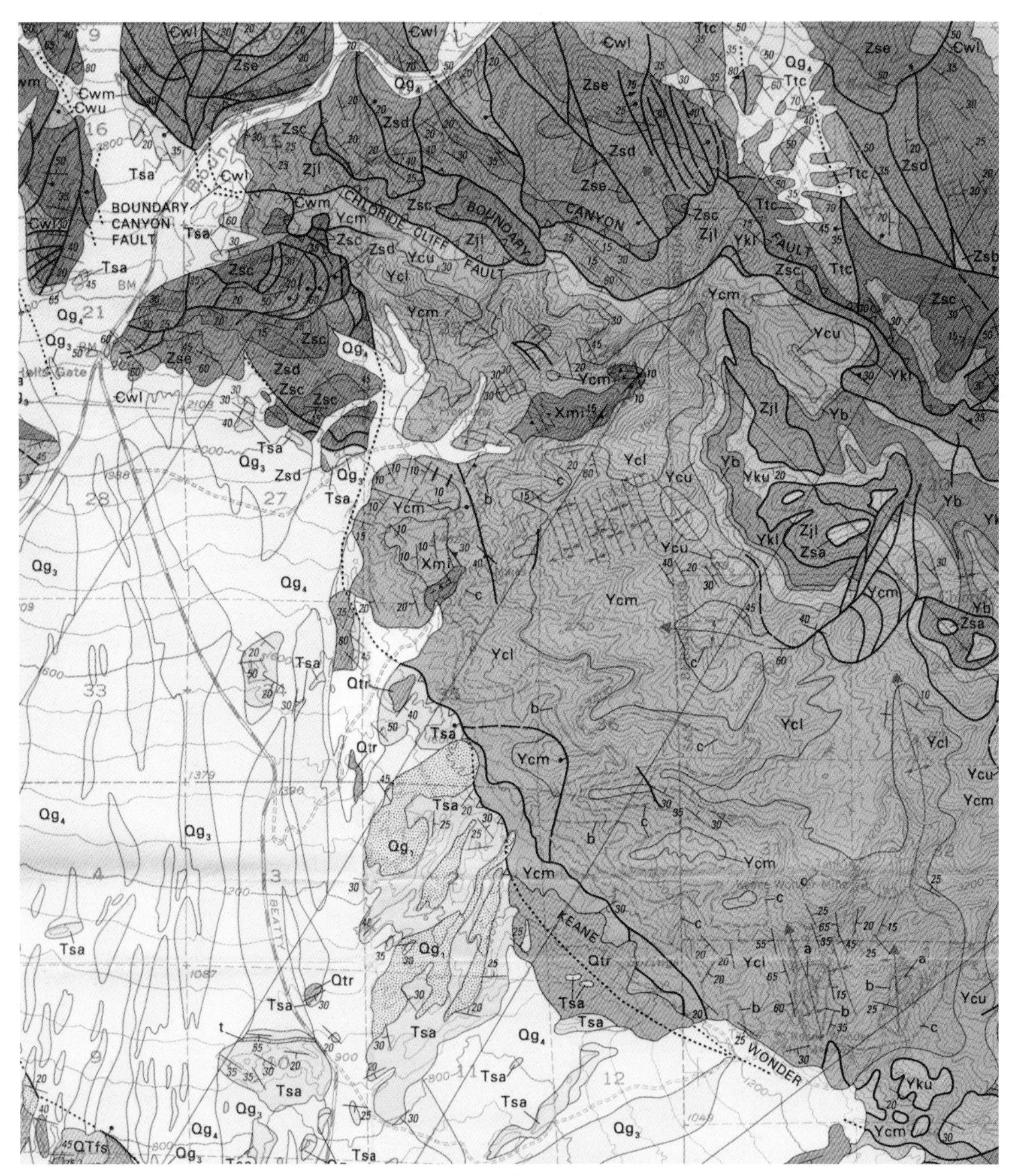

Map 2. Bedrock geologic map of the front of the Funeral Mountains north of, and including the Keane Wonder Mine, from Wright and Troxel, 1993. North is to the top; map area measures approximately six miles from north to south. Location "K" is the Keane Wonder Mine; location "M" is the head of Monarch Canyon. Abbreviations of the most prevalent rock units are as follows: Xmi: Precambrian basement rock; Ycl (gray), Ycm (blue), Ycu (light blue): Lower, Middle, and Upper Crystal Spring Formation; Yb (blue): Beck Springs Dolomite; Ykl (gray) and Yku (light gray): Lower and Upper Kingston Peak Formation; Zjl (olive) Lower Johnnie Formation: Zsc (dark brown), Zsd (light brown), Zse (brown): Stirling Formation; Cwl (brown): Wood Canyon Formation; Ttc (orange): Titus Canyon Formation.

Appendices

Dust Devil and Death Valley salt pan, view towards the west.

Glossary

Alluvial fans. Fan-shaped landforms, found typically at the bases of mountains, of material eroded from the mountains. Most strictly, this material consists of alluvium, which is water-transported, but in many cases it also consists of material that was transported primarily by gravity.

Alluvium. Water-transported material.

Andesitic. A compositional term used to describe magmas or volcanic rocks that consist of between 60–65% silica.

Anticline. A fold in layered rock in which the oldest rock lies in the core. The beds on either side of an anticline generally tilt away from the core.

Arkosic. A term used to describe sedimentary rocks that contain abundant grains of feldspar.

Bajada. Coalesced alluvial fans along the front of a mountain range.

Basaltic. A compositional term used to describe magmas or volcanic rocks that consist of between 50–55% silica. Basaltic rocks are typically dark gray to black in color. Generally, rocks with 55–60% silica are called basaltic andesites.

Breccia. A type of rock with angular particles that exceed 2 mm in size. Breccias may be sedimentary in origin, or form by crushing within fault zones.

Chaos. A structural term for a mosaic of fault-bounded, typically gigantic blocks, derived from a stratigraphic succession and arranged in proper stratigraphic order, but occupying only a small fraction of the thickness of the original succession. In the Death Valley region, where L.F. Noble coined the term, chaos in normally viewed as a product of extreme crustal extension.

Clasts. Individual rock or mineral fragments in a sedimentary rock.

Conglomeratic. A term used to describe a conglomerate: a sedimentary rock with particle sizes that exceed 2 mm.

Crystalline complex. A body of metamorphic and igneous rock that forms the oldest and originally deepest part of a region.

Dacite. A compositional term used to describe volcanic rocks that consist of between 65–70% silica. Dacites generally resemble andesites, but generally contain less-calcic plagioclase and more quartz.

Desert Varnish. Black staining on rock, consisting of manganese oxide.

Detritus. Debris.

Diabase. A term used to describe a rock that is basaltic in composition, but crystallized at a shallow level in the crust rather than erupted on the surface and cooled as a lava flow.

Diamictite. Sedimentary accumulations of highly variably sized particles set in a muddy matrix. Many diamictites are interpreted as ancient glacial deposits.

Dikes. Tabular-shaped intrusions that cut across layering in the host rock.

Diorite. A compositional term used to describe intrusive igneous rocks that consist of between 60–70% silica. Diorites are chemically equivalent to andesites.

Ephemeral Lakes. Lakes that are filled with water for only short periods of time.

Extensional tectonics. The study of regions in which the deformation of the earth's crust has been primarily extensional.

Faceted spurs. An abrupt termination of a ridge at the front of a mountain range that was caused by displacement along a fault.

Fault scarp. A sudden rise or step in the landscape that formed by slip on a fault zone.

Fault-block range. A mountain range that has risen along a master fault zone. Most ranges in Death Valley are tilted fault blocks, because they have risen and tilted along normal faults.

Faults. Breaks or fractures in rock along which one side has moved relative to the other. The type of relative movement determines the type of fault (see Basics About Faults and Fault Zones).

- **Detachment fault.** A type of low-angle normal fault in which deformation of the block above the fault surface occurred independent of the deformation below the fault.

- **Dip-slip faults.** Faults in which movement occurred parallel to the dip, or direction of inclination, of the fault plane.

- **Normal fault.** A type of dip slip fault in which the block of rock beneath the fault surface (the footwall) has moved upwards relative to the block above the fault surface (hanging wall). High-angle normal faults are those in which the fault surface is inclined at an angle of greater than 45°. Low-angle normal faults are those in which the fault surface is inclined at an angle of less than 30°.

- **Strike-slip faults.** Faults in which movement occurred parallel to the strike, or horizontal direction, of the fault plane.

Thrust fault. A type of dip slip fault in which the block of rock beneath the fault surface (the footwall) has moved downwards relative to the block above the fault surface (hanging wall). Thrust faults typically bring older rocks over the top of younger rocks.

Felsic. A term used to describe igneous rocks that contain abundant feldspar.

Fluvial deposits. Sedimentary accumulations, deposited by streams or rivers.

Gabbro. A compositional term used to describe intrusive igneous rocks that consist of between 50–60% silica. Gabbros are chemically equivalent to basalts. They typically dark in color and contain no quartz.

Igneous. A type of rock that formed by the cooling and crystallization of magma (liquid rock).

Interdune areas. Patches of bare ground between sand dunes.

Mafic. A term used to describe igneous rocks that are generally low in silica.

Magma. Liquid rock that is beneath the earth's surface.

Megabreccia. A very coarse, large-scale breccia.

Microbial mat. Literally, a mat of microbes. Such mats entrap fine-grained carbonate sediment that through time forms a stack of thin wavy layers of limestone. Their fossil forms are called stromatalites.

Mylonites. Fine-grained metamorphic rocks that display a strong foliation and lineation. The fine grain size results from extreme deformation under hot conditions.

Passive Continental margin. A continental margin, such as the east coast of North America, that does not border a convergent plate margin.

Pegmatite. A type of silica-rich intrusive igneous rock with very coarse-grained feldspar.

Perennial Lakes. Lakes that are constantly filled with water.

Playa. An exceedingly flat mud-covered surface that occasionally becomes filled with water.

Plutons. Bodies of intrusive rock.

Rain Shadow. An area on the downwind side of a mountain range. Rain shadows are especially arid because the winds lose their moisture as they rise over the mountains.

Rhyolitic. A compositional term used to describe magmas or volcanic rocks that consist of greater than 70% silica. Rhyolitic rocks tend to be light in color.

Salt pan. An exceedingly flat salt-covered surface.

Sand Dunes. Large accumulations of wind-blown sand.

Crescentic dune. A type of sand dune with two arms that is asymmetrical in cross-section.

Linear dune. A type of sand dune with two arms that is symmetrical in cross-section.

Star dune. A type of sand dune with at least three arms.

Sills. Tabular-shaped intrusions that are parallel to layering in the host rock.

Stromatalites. The fossil form of a microbial mat.

Syncline. A fold in layered rock in which the youngest rock lies in the core. The beds on either side of a syncline generally tilt towards from the core.

Tufa. Calcium carbonate that is precipitated in lakes.

Venitfacts. Wind-faceted rocks.

Wineglass Canyon. A canyon whose shape resembles a wineglass, with a narrow, steep mouth and wider, flaring walls towards the back. The alluvial fan at the front resembles the base; the mouth resembles the stem; the wider up-canyon region resembles the bowl.

References

Geologic Maps

Albee, A. L., Labotka, T. C., Lanphere, M. A., and McDowell, S. D., 1981, Geologic map of the Telescope Peak Quadrangle, California: U. S. Geological Survey Geologic Quadrangle Map GQ-1532, scale 1: 62,500.

Burchfiel, B.C., 1969, Geology of the Dry Mountain Quadrangle, Inyo County, California: California Division of Mines and Geology Special Report 99, v. 19.

Chesterman, C. W., 1973, Geology of the northeast quarter of the Shoshone Quadrangle, Inyo County, California: California Division of Mines and Geology Map Sheet 18, 1:24,000.

Drewes, H., 1963, Geology of the Funeral Peak Quadrangle, California, on the eastern flank of Death Valley: U. S. Geological Survey Professional Paper 413, 78 p., scale 1:48.000.

Greene, R.C., 1997, Geology of the northern Black Mountains, Death Valley, California: U.S. Geological Survey Open File Report OF 97-79, 110 p., 1:24,000.

Hall, W. E., 1971, Geology of the Panamint Butte quadrangle, Inyo County, California: U. S. Geological Survey Bulletin 1299, 67 p., scale 1:48.000.

Jennings, C. W., Burnett, J. L., and Troxel, B. W., 1962, Trona sheet: California Division of Mines and Geology Geologic Atlas of California, scale 1:250,000.

McAllister, J.F., 1956, Geology of the Ubehebe Peak Quadrangle, California: U.S. Geological Survey Geologic Quadrangle map GQ 95, 1:62,500.

McAllister, J. F., 1970, Geology of the Furnace Creek borate area, Inyo County, California: California Division of Mines Map Sheet 14, scale 1:24,000.

McAllister, J. F., 1971, Preliminary geologic map of the Funeral Mountains in the Ryan quadrangle, Inyo County, California: U. S. Geological Survey Open File Map, scale 1: 31,680.

McAllister, J. F., 1973, Geologic map of the Amargosa Valley borate area,—southeast continuation of the Furnace Creek area: U. S. Geological Survey Miscellaneous Geologic Investigations Map 1-782, scale 1: 31,680.

Snyder, C.T., Hardman, G., and Zdenek, F.F., 1964, Pleistocene Lakes in the Great Basin: United States Geological Survey Miscellaneous Geologic Investigations Map I-416, scale 1:1,000,000.

Streitz, R., and Stinson, M. C., 1977, Death Valley sheet: California Division of Mines and Geology Geologic Atlas of California, scale 1:250,000.

Strand, R.G., 1967, Geologic map of California, Mariposa Sheet, California Division of Mines and Geology, 1:250,000.

Thelin, G.P., and Pike, R.J., 1991, Landforms of the Conterminous United States—A Digital Shaded-Relief Portrayal: United States Geological Survey Miscellaneous Investigations Map, I-2206.

Wright, L. A., and Troxel, B. W., 1984, Geology of the north 1/2 Confidence Hills 15' quadrangle, Inyo County, California: California Division of Mines and Geology Map Sheet 34, scale 1:24,000.

Wright, L. A., Troxel, B. W., Burchfiel, B. C., Chapman, R., and Labotka, T., 1981, Geologic cross section from the Sierra Nevada to the Las Vegas Valley, eastern California to western Nevada: Geological Society of America Map and Chart Series, MC-28M, scale 1:250,000.

Wright, L. A., and Troxel, B. W., 1993. Geologic map of the central and northern Funeral Mountains and adjacent areas, Death Valley region, southern California: U. S. Geological Survey Miscellaneous Investigations Series, Map I-2305, scale 1:48,000.

Literature References

Applegate, J. D. R., Walker, J. D., and Hodges, K. V., 1992, Late Cretaceous extensional unroofing in the Funeral Mountains metamorphic core complex, California: Geology, v. 20, p. 519–522.

Bacon, D., Cahill, T., and Tombrello, T.A., 1996, Sailing stones on Racetrack Playa: Journal of Geology, v. 104, p. 121–125.

Blackwelder, E., 1933, Lake Manly—An extinct lake of Death Valley: Geographical Review, v. 23, p. 464–471.

Blakely, R.J., Jachens, R.C., Calzia, J.P., and Langenheim, V.E., 1999, Cenozoic basins of the Death Valley extended terrane as reflected in regional-scale gravity anomalies, in Wright, L.A., and Troxel, B.W., eds., Cenozoic Basins of the Death Valley Region: Boulder, Colorado, Geological Society of America Special Paper 333, p. 1–16.

Brogan, G.E., Kellogg, K.S., Slemmons, D.B., and Terhune, C.L., 1991, Late Quaternary faulting along the Death Valley-Furnace Creek fault system, California and Nevada: U.S. Geological Survey Bulletin 1991, 23p + maps.

Brown, J.H., 1971, The desert pupfish: Scientific American, v. 225, p. 104–110.

Burchfiel, B. C., and Stewart, J. H., 1966, "Pull-apart" origin of the central segment of Death Valley, California: Geological Society of America Bulletin, v. 77, p. 439–432.

Burchfiel, B.C., Hodges, K.V., and Royden, L.H., 1987, Geology of Panamint Valley—Saline Valley pull-apart system, California: Palinspastic evidence for low-angle geometry of a Neogene range-bounding fault: Journal of Geophysical Research, v. 92, p. 10,422–10,426.

Butler, P. R., Troxel, B.W., and Verosub, K. L., 1988, Late Cenozoic history and styles of deformation along the southern Death Valley fault zone, California: Geological Society of America Bulletin, v. 100, p. 402–410.

Cemen, I., and Wright, L.A., 1990, Effect of Cenozoic extension on Mesozoic thrust surfaces in the central and southern Funeral Mountains, Death Valley, California:

Cemen, I., Wright, L. A., Drake, R. E., and Johnson, F. C., 1985, Cenozoic sedimentation and sequence of deformational events at the southeastern end of the Furnace Creek strike-slip fault zone, Death Valley region, California, *in* Biddle, K. T., and Christie-Blick, Nicholas, eds., Strike-slip deformation and basin formation: Society of Economic Paleontologists and Mineralogists Special Publication, No. 37, p. 127–141.

Cichanski, M.A., 1990, Stratigraphy and structure of the upper plate of the Badwater turtleneck, Death Valley, California (senior thesis): Seattle, University of Washington, 23p.

Corsetti, F. A., and Hagadorn, J. W., 2000, Precambrian-Cambrian transition: Death Valley, United States: Geology, v. 28, p. 299–302.

Cowan, D.S., Cladouhos, T.T., Morgan, J.K., 2003, Structural geology and kinematic history of rocks formed along low-angle normal faults, Death Valley, California: Geological Society of America Bulletin, v. 115, p. 1230–1248.

Crowe, B.M., and Fisher, R.V., 1973, Sedimentary structures in base-surge deposits with special reference to cross bedding, Ubehebe Crater, Death Valley, California: Geological Society of America Bulletin, v. 84, p. 663–682.

Curry, H.D., 1938, "Turtleback" fault surfaces in Death Valley, California: Geological Society of America Bulletin, v. 49, p. 1875.

De Voogd, B., Serpa, L., Brown, L., Hauser, E., Kaufman, S., Oliver, J., Troxel, B., Willemin, B.J., and Wright, L.A., 1986, The Death Valley bright spot: A midcrustal magma body in the southern Great Basin, California?: Geology, v. 14, p. 64–67.

DeWitt, E. H., Armstrong, R. L., Sutter, J. F., and Zartman, R. E., 1984, U-Th-Pb, Rb-Sr, and Ar-Ar mineral and whole-rock isotopic systematics in a metamorphosed granitic terrane, southeastern California: Geological Society of America Bulletin, v. 95, p. 723–739.

Diehl, P., 1976, Stratigraphy and sedimentology of the Wood Canyon Formation, Death Valley area, California: *in* Troxel, B. W., and Wright, L.A. eds., Geologic Features: Death Valley region, California. California Division of Mines and Geology Special Report 106, p. 51–62.

Evans, J.R., Taylor, G.C., and Rapp, J.S., 1976, Mines and mineral deposits in Death Valley National Monument, California: California Division of Mines and Geology Special Report 125, 60p.

Haefner, Richard, 1976, Geology of the Shoshone Volcanics, Death Valley region, eastern California: *in*, Troxel, B.W., and Wright, L.A., eds., Geologic Features, Death Valley, California, California Division of Mines and Geology, Special Report 106, p. 67–72.

Haefner, Richard, and Troxel, Bennie, 2002, A petrologic paradox in central Death Valley, California (Abs.): Geological Society of America Abstracts with Programs, v. 34, no. 4, p. 8.

Hayman, N.W., Knott, J.R., Cowan, D.S., Nemser, E., and Sarna-Wojcicki, A.M., Quaternary low-angle slip on detachment faults in Death Valley, California: Geology, v. 31, p. 343–346.

Heaman, L. M., Grotzinger, J. P. , 1992, 1.08 Ga diabase sills in the Pahrump Group, California; implications for development of the Cordilleran Miogeocline: Geology, v. 20, p. 637–640.

Hodges, K. V., McKenna, L. W., Stock, J., Knapp, L., Page, L., Sterniof, K., Silverberg, D., Wust, G., and Walker, J. D., 1989, Evolution of extensional basins and Basin and Range topography west of Death Valley, California: Tectonics, v. 8, p. 453–467.

Hodges, K. V., and Walker, J.D., 1992, Extension in the Cretaceous Sevier orogen, North American Cordillera: Geological Society of America Bulletin, v. 104, p. 560–569.

Hoffman, P.F., and Schrag, D.P., 2000, Snowball Earth: Scientific American, v. 282, p. 68–75.

Hoisch, T.D., and Simpson, C., 1993, Rise and tilt of metamorphic rocks in the lower plate of a detachment fault in the Funeral Mountains, Death Valley, California: Journal of Geophysical Research, v. 98, p. 6805–6827.

Holm, D. K., Snow, J. K., and Lux, D. R., 1992, Thermal and barometric constraints on the intrusion and unroofing history of the Black Mountains, Death Valley, CA: Tectonics, v. 11, p. 507–522.

Hooke, R.L., 1972, Geomorphic evidence for Late-Wisconsin and Holocene tectonic deformation, Death Valley, California: Geological Society of America Bulletin, v. 83, p. 2073–2098.

Horodyski, R. J., Gehling, J. G., Jensen, S., and Runnegar, B., 1994, Ediacara fauna and earliest Cambrian trace fossils in a single parasequence set, southern Nevada (Abs): Geological Society of America Abstracts with Programs, v. 26, p. 60.

Hunt, C. B., and Mabey, D. R., 1966, Stratigraphy and structure, Death Valley, California: U. S. Geological Survey Professional Paper 494A, 162 p.

Hunt, C. B., 1975, Death Valley: geology, ecology, archeology: University of California Press, Berkeley and Los Angeles, California, 234 p.

Klinger, R.E., 1999, Tectonic geomorphology along the Death Valley fault system—Evidence for recurrent Late Quaternary activity in Death Valley National Park: in Slate, J.L., ed., Proceedings of Conference on Status of Geologic Research and Mapping, Death Valley National Park, U.S. Geological Survey Open-File Report 99-153, p. 132–139.

Klinger, R. E., 2001a, Late Quaternary volcanism of Ubehebe Crater, in Machette, M.N., Johnson, M.L., and Slate, J.L., eds., Quaternary and Late Pliocene Geology of the Death Valley region: Recent observations on tectonics, stratigraphy, and lake cycles. Guidebook for the 2001 Pacific Cell—Friends of the Pleistocene Fieldtrip. United States Geological Survey Open File Report 01-51, p. A21–A24.

Klinger, R. E., 2001b, Northern Death Valley, Road Log for Day A, in Machette, M.N., Johnson, M.L., and Slate, J.L., eds., Quaternary and Late Pliocene Geology of the Death Valley region: Recent observations on tectonics, stratigraphy, and lake cycles. Guidebook for the 2001 Pacific Cell—Friends of the Pleistocene Fieldtrip. United States Geological Survey Open File Report 01–51, p. A7–A20.

Klinger, R. E., Piety, L.A., and Machette, M.A., 2001, Late Quaternary growth of the Echo Canyon thrust and Texas Springs syncline, in Machette, M.N., Johnson, M.L., and Slate, J.L., eds., Quaternary and Late Pliocene Geology of the Death Valley region: Recent observations on tectonics, stratigraphy, and lake cycles. Guidebook for the 2001 Pacific Cell—Friends of the Pleistocene Fieldtrip. United States Geological Survey Open File Report 01-51, p. B75–B79. Open File Report available on the web at http://pubs.usgs.gov/of/2001/ofr-01-0051/.

Knott, J.R., and Wells, S.G., 2001, Late Pleistocene Slip rate of the Black Mountains fault zone, in Machette, M.N., Johnson, M.L., and Slate, J.L., eds., Quaternary and Late Pliocene Geology of the Death Valley region: Recent observations on tectonics, stratigraphy, and lake cycles. Guidebook for the 2001 Pacific Cell—Friends of the Pleistocene Fieldtrip. United States Geological Survey Open File Report 01-51, p. C103–C104. Open File Report available on the web at http://pubs.usgs.gov/of/2001/ofr-01-0051/.

Knott, J.R., 1999, Quaternary stratigraphy and geomorphology of Death Valley: *in* Slate, J.L., ed., Proceedings of conference on status of geologic research and mapping, Death Valley National Park, U. S. Geological Survey Open-File Report 99-153, p. 90–96.

Knott, J.R., Sarna-Wojcicki, A.M., Meyer, C.E., Tinsley, J.C., III, Wells, S.G., and Wan, E., 1999, Late Cenozoic stratigraphy and tephrochronology of the western Black Mountains piedmont, Death Valley, California: Implications for the tectonic development of Death Valley, in Wright, L.A., and Troxel, B.W., ed., Cenozoic basins of the Death Valley region: Geological Society of America Special Paper 333, p. 345–366.

Labotka, T.C., and Albee, A.L., 1988, Metamorphism and tectonics of the Death Valley region, California and Nevada, in Ernst, W.G., ed., Metamorphism and Crustal Evolution of the Western United States: Englewood Cliffs, Prentice-Hall, p. 714–736.

Lingenfelter, R.E., 1986, Death Valley and the Amargosa. University of California Press, Berkeley, 664p.

Lowenstein, T.K., Li, L., Brown, C., Roberts, S.M., Ku, T., Luo, S., and Yang, W., 1999, 200 k.y. paleoclimate record from Death Valley salt core: Geology, v. 27, p. 3–6.

Machette, M.N., and Crone, A.J., 2001, Late Holocene faulting on the Old Ghost alluvial-fan complex: *in* Machette, M.N., Johnson, M.L., and Slate, J.L., eds., Quaternary and Late Pliocene Geology of the Death valley region: Recent observations on tectonics, stratigraphy, and lake cycles. Guidebook for the 2001 Pacific Cell—Friends of the Pleistocene Fieldtrip. United States Geological Survey Open File Report 01-51, p. B67–B75. Open File Report available on the web at http://pubs.usgs.gov/of/2001/ofr-01-0051/.

Machette, M.N., Klinger, R.E., Knott, J.R., Wills, C.J., Bryant, W.A., and Reheis, M.C., 2001a, A proposed nomenclature for the Death Valley fault system: in Machette, M.N., Johnson, M.L., and Slate, J.L., eds., Quaternary and Late Pliocene Geology of the Death Valley region: Recent observations on tectonics, stratigraphy, and lake cycles. Guidebook for the 2001 Pacific Cell—Friends of the Pleistocene Fieldtrip. United States Geological Survey Open File Report 01–51, p. J173–J183.

Machette, M.N., Klinger, R.E., and Knott, J.R., 2001b, Questions about Lake Manly's age, extent, and source: in Machette, M.N., Johnson, M.L., and Slate, J.L., eds., Quaternary and Late Pliocene Geology of the Death Valley region: Recent observations on tectonics, stratigraphy, and lake cycles. Guidebook for the 2001 Pacific Cell—Friends of the Pleistocene Fieldtrip. United States Geological Survey Open File Report 01–51, p. G143–G149. Open File Report available on the web at http://pubs.usgs.gov/of/2001/ofr-01-0051/.

Machette, M. N., and Slate, J.L., 2001, Late Quaternary uplift of the Mustard Canyon hills—tectonic, diapiric, or both?: *in* Machette, M.N., Johnson, M.L., and Slate, J.L., eds., Quaternary and Late Pliocene Geology of the Death Valley region: Recent observations on tectonics, stratigraphy, and lake cycles. Guidebook for the 2001 Pacific Cell—Friends of the Pleistocene Fieldtrip. United States Geological Survey Open File Report 01-51, p. B59–B63. Open File Report available on the web at http://pubs.usgs.gov/of/2001/ofr-01-0051/.

McAlllister, J., 1976, Geologic maps and sections of a strip from Pyramid Peak to the southeast end of the Funeral Mountains, Ryan Quadrangle, California: in, Troxel, B.W., and Wright, L.A., eds., Geologic Features, Death Valley, California, California Division of Mines and Geology, Special Report 106, p. 63–66.

Meek, N., and Dorn, R., 2000, Is Mushroom Rock a ventifact?: California Geology, v. 53, p. 18–21.

Messina, P. and Stoffer, P., 2000, Terrain analysis of the Racetrack Basin and the sliding rocks of Death Valley: Geomorphology, v. 35, p. 253–265.

Messina, P. and Stoffer, P., 2001, Using new technology to solve an old mystery: California Geology, v. 54, n. 1, p. 4–15.

Miller, G.A., 1977, Appraisal of the water resources of Death Valley, California-Nevada: U.S. Geological Survey Open-File Report 77-728, 68 p.

Miller, M., 1991, High-angle origin of the currently low-angle Badwater Turtleback fault, Death Valley, California: Geology, v. 19, p. 372–375.

Miller, M., 1996, Ductility in fault gouge from a normal fault system, Death Valley, California: A mechanism for fault-zone strengthening and relevance to paleoseismicity: Geology, v. 24, p. 603–606.

Miller, M., 1999a, Implications of ductile strain on the Badwater Turtleback for pre-14 Ma extension in the Death Valley region, California: *in* Wright, L.A., and Troxel, B.W. eds., Tertiary basins in the Death Valley region, California: Geological Society of America Special Paper 333, p. 115–126.

Miller, M., 1999b, Bedrock geology near Badwater Spring: *in* Slate, J.L., ed., Proceedings of Conference on Status of Geologic Research and Mapping, Death Valley National Park, U.S. Geological Survey Open-File Report 99-153, p. 170–171.

Miller, M., 2000, Probable basement-involved thrust faulting in the Black Mountains turtlebacks, Death Valley, California (Abs.): Geological Society of America Abstracts with Programs, v. 32, p. 46.

Miller, M., 2001, Death Valley's visible history: a new geologic map and accompanying photographs: California Geology, v. 54, no. 2, p. 4–15.

Miller, M., 2003, Basement-involved thrust faulting in a thin-skinned fold-and-thrust belt, Death Valley, California, USA: Geology, v. 31, p. 31–34.

Miller, M., and Friedman, R., 2003, New U-Pb zircon ages indicate major extension in Death Valley, CA predated 9.5 Ma: implications for models of crustal extension: Geological Society of America Abstracts with Programs.

Miller, M., and Pavlis, T. L., 2003, The Black Mountains Turtlebacks: Rosetta Stones of Death Valley geology. Special issue of Earth Science Reviews, Ed. James Calzia, Elsevier.

Miller, M., and Prave, A.R., 2002, Rolling hinge or fixed basin?: A test of continental extensional models in Death Valley, California, United States: Geology, v. 30, p. 847–850.

Miller, R.R., 1950, Speciation in fishes of the genera *Cyprinodon* and *Empetrichthys*, inhabiting the Death Valley region: Evolution, v. 4, p. 155–163.

Nemser, E.S., 2001, Kinematic development of upper plate faults above low-angle normal faults in Death Valley, California (M.S. thesis): Seattle, University of Washington, 102p.

Niemi, N.A., Wernicke, B.P., Brady, R.J., Saleeby, J.B., and Dunne, G.C., 2001, Distribution and provenance of the middle Miocene Eagle Mountain Formation, and implications for regional kinematic analysis of the Basic and Range Province: Geological Society of America Bulletin, v. 113, p. 419–442.

Niemi, N.A., 2002, Extensional tectonics in the Basin and Range Province and the geology of the Grapevine Mountains, Death Valley region, California and Nevada [Ph.D. thesis]: Pasadena, California Institute of Technology, 364 p.

Noble, L. F., 1941, Structural features of the Virgin Spring area, Death Valley, California: Geological Society of America Bulletin: v. 52, p. 941–1000.

Otton, J.K., 1976, Geologic features of the central Black Mountains, Death Valley, California: *in* Troxel, B. W., and Wright, L.A. eds., Geologic Features: Death Valley region, California. California Division of Mines and Geology Special Report 106, p. 27–34.

Pavlis, T.L., Serpa, L.F., and Keener, C., 1993, Role of seismogenic processes in fault-rock development: an example from Death Valley, California: Geology, v. 21, p. 267–270.

Reid, J.B., Bucklin, E.P., Copenagle, L., Kidder, J., Pack, S.M., Polissar, P.J., and Williams, M.L., 1995, Sliding rocks at the Racetrack, Death Valley: What makes them move?: Geology, v. 23, p. 819–822.

Renik, B., and Christie-Blick, N., 2004, Re-evaluation of Miocene evidence for extreme crustal extension across the Death Valley Region, Calfironia: American Association of Petroleum Geologists annual meeting.

Reynolds, M.W., 1969, Stratigraphy and structural geology of the Titus and Titanothere canyons area, Death Valley, California: Berkeley, California, University of California Ph.D. dissertation, 310p.

Reynolds, M. W., 1976, Geology of the Grapevine Mountains, Death Valley, California: A summary, *in* Troxel, B. W., and Wright, L.A. eds., Geologic Features: Death Valley region, California. California Division of Mines and Geology Special Report 106, p. 19–26.

Reynolds, M.W., Wright, L.A., and Troxel, B.W., 1996, Evidence for Tertiary age of recumbent folds, Grapevine and northern Funeral Mountains, Death Valley, California (Abs): Geological Society of America Abstracts with Programs, v. 28, p. 513.

Roberts, M.T., 1976, Stratigraphy and depositional environments of the Crystal Spring Formation, southern Death Valley region, California: *in* Troxel, B. W., and Wright, L.A. eds., Geologic Features: Death Valley region, California. California Division of Mines and Geology Special Report 106, p. 35–44.

Saylor, B. Z., and Hodges, K. V., 1991, The Titus Canyon Formation; evidence for early Oligocene extension in the Death Valley area, CA (Abs.): Geological Society of America, Abstracts with Programs, v. 23, no. 5, p. 82.

Serpa, L., de Voogd, B., Wright, L., Willemin, J., Oliver, J., Hauser, E., and Troxel, B., 1988, Structure of the central Death Valley pull-apart basin and vicinity from COCORP profiles in the southern Great Basin: Geological Society of America Bulletin, v. 100, p. 1437–1450.

Serpa, L.F., and Pavlis, T. L., 1996, Three-dimensional model of the Cenozoic history of the Death Valley region, southeastern California: Tectonics, v. 15, p. 1113–1128.

Sharp, R.P., and Carey, D.L., 1976, Sliding stones, Racetrack Playa, California: Geological Society of America Bulletin, v. 87, p. 1704–1717.

Sharp, R.P. and Glazner, A.F., 1997, Geology Underfoot in Death Valley and Owens Valley. Mountain Press Publishing, Missoula, Montana. 321 p.

Snow, J.K., 1990, cordilleran orogenesis, extensional tectonics and geology of the Cottonwood Mountains area, Death Valley region, California and Nevada [Ph.D. thesis]: Cambridge, Harvard University, 533 p.

Snow, J. K., 1992, Paleogeographic and structural significance of an Upper Mississippian facies boundary in southern Nevada and east-central California: Discussion: Geological Society of America Bulletin, v. 104, p. 1067–1069.

Snow, J.K., and Wernicke, B.W., 1989, Uniqueness of geological correlations: An example form the Death Valley extended terrain: Geological Society of America Bulletin, v. 101, p. 1351–1362.

Snow, J.K., Asmerom, Y., and Lux, D.R., 1991, Permian-Triassic plutonism and tectonics, Death Valley region, California and Nevada: Geology, v. 19, p. 629–632.

Snow, J.K., and Lux, D.R., 1999, Tectono-sequence stratigraphy of Tertiary rocks in the Cottonwood Mountains and northern Death Valley area, California and Nevada: *in* Wright, L.A., and Troxel, B.W. eds., Tertiary basins in the Death Valley region, California: Geological Society of America Special Paper 333, p. 17–64.

Snow, J.K., and Wernicke, B.W., 2000, Cenozoic tectonism in the central Basin and Range: magnitude, rate, and distribution of upper crustal strain: American Journal of Science, v. 300, p. 659–719.

Soltz, D.L., and Naiman, 1978, The natural history of native fishes in the Death Valley system: Natural History Museum of Los Angeles County, in conjunction with the Death Valley Natural History Association, Science Series 30, 76p.

Stanley, G.M., 1955, Origin of playa stone tracks, Racetrack Playa, Inyo County, California: Geological Society of America Bulletin, v. 66, p. 1329–1350.

Steinkampf, W.C., and Werrell, W.L., 2001, Ground-Water flow to Death Valley, as inferred from the chemistry and geohydrology of selected springs in Death Valley National Park, California and Nevada: US Geological Survey Water-Resources Investigations Report 98-4114, 37p.

Stevens, C. H., Stone, P., and Belasky, P., 1991, Paleogeographic and structural significance of an Upper Mississippian facies boundary in southern Nevada and east-central California: Geological Society of America Bulletin, v. 103, p. 876–885.

Stevens, C. H., Stone, P., and Belasky, P., 1992, Paleogeographic and structural significance of an Upper Mississippian facies boundary in southern Nevada and east-central California: Reply: Geological Society of America Bulletin, v. 104, p. 1069–1071.

Stewart, J. H., Albers, J. P., and Poole, F. G., 1968, Summary of regional evidence for right-lateral displacement in the western Great Basin: Geological Society of America Bulletin, v. 79, p. 1407–1414.

Stewart, J.H., 1983, Extensional tectonics in the Death Valley area, California: Transport of the Panamint Range structural block 80 km northwestward: Geology, v. 11, p. 153–157.

Topping, David, J., 1993, Paleogeographic reconstruction of the Death Valley extended region: Evidence from Miocene large rock-avalanche deposits in the Amargosa Chaos basin, California: Geological Society of America Bulletin, v. 105, p. 1190–1213.

Topping, David, J., 2003, Stratigraphic constraints on the style and magnitude of extension in the southern Black Mountains, Death Valley, California—10.5Ma to present. Geological Society of America Abstracts with Programs, v. 35, p. 347.

Troxel, B. W., 1974, Significance of a man-made diversion of Furnace Creek Wash at Zabriskie Point, Death Valley, California: in Guidebook: Death Valley region, California and Nevada, Death Valley Publishing Co., Shoshone, CA, p. 87–91.

Troxel, B.W., 1974, Geologic guide to the Death Valley region, California and Nevada: *in* Wright, L.A., and Troxel, B.W., eds, Guidebook: Death Valley region, California and Nevada, Geological Society of America, Boulder, p. 2–16.

Troxel, B.W., and Heydari, E., 1982, Basin and Range geology in a roadcut: in Cooper, J.D., Troxel, B.W., and Wright, L.A., eds. Geology of selected areas in the San Bernardino Mountains, Western Mojave Desert, and southern Great Basin, California: Geological Soc. of America volume and guidebook, Anaheim, p. 91–96.

Wasserburg, G.J.F., Wetherill, G.W., and Wright, L.A., 1959, Ages in the Precambrian terrane of Death Valley, California: Journal of Geology, v. 67, p. 702–708.

Wernicke, B., Axen, G. C., and Snow, J. K., 1988, Basin and Range extensional tectonics at the latitude of Las Vegas, Nevada: Geological Society of America Bulletin, v. 100, p. 1738–1757.

Wernicke, B.P., Snow, J.K., Axen, G.J., Burchfiel, B.C., Hodges, K.V., Walker, J.D., and Guth, P.L., 1989, Extensional tectonics in the Basin and Range Province between the southern Sierra Nevada and the Colorado Plateau: American Geophysical Union Field Trip Guidebook T138.

Whitney, D.L., Hirschmann, M., and Miller, M.G., 1993, Zincian ilmenite—Ecandrewsite from a pelitic schist, Death Valley, California, and the paragenesis of $(Zn,Fe)Ti0_3$ solid solution in metamorphic rocks: Canadian Mineralogist, v. 31, p. 425–436.

Williams, E.G., Wright, L.A., and Troxel, B.W., 1976, The Noonday Dolomite and equivalent stratigraphic units, southern Death Valley region, California: *in* Troxel, B. W., and Wright, L.A. eds., Geologic Features: Death Valley region, California. California Division of Mines and Geology Special Report 106, p. 45–49.

Wills, C.J., 2001, Liquefaction in the California Desert—an unexpected geologic hazard, in Machette, M.N., Johnson, M.L., and Slate, J.L., eds., Quaternary and Late Pliocene Geology of the Death Valley region: Recent observations on tectonics, stratigraphy, and lake cycles. Guidebook for the 2001 Pacific Cell—Friends of the Pleistocene Fieldtrip. United States Geological Survey Open File Report 01-51, p. O225–O231.

Wright, L.A., 1968, Talc deposits of the southern Death Valley—Kingston Range region, California. California Division of Mines and Geology, Special Report 95, 79p.

Wright, L.A., 1989, Overview of the role of strike-slip and normal faulting in the Neogene history of the region northeast of Death Valley, California—Nevada: *in* Late Cenozoic evolution of the Southern Great Basin, Open File 89-1, edited by M. Ellis, p. 1–12.

Wright, L.A., Otton, J.K., and Troxel, B.W., 1974, Turtleback surfaces of Death Valley viewed as phenomena of extensional tectonics: Geology, v. 2, p. 53–54.

Wright, L.A., Troxel, B.W., Williams, E.G., Roberts, M.T., and Diehl, P.E., 1976, Precambrian sedimentary environments of the Death Valley region, eastern California: *in* Troxel, B. W., and Wright, L.A. eds., Geologic Features: Death Valley region, California. California Division of Mines and Geology Special Report 106, p. 7–15.

Wright, L.A., Williams, E.G., and Cloud, P., 1976, Stratigraphic cross-section of Proterozoic Noonday Dolomite, War Eagle Mine area, southern Nopah Range, eastern California: *in* Troxel, B. W., and Wright, L.A. eds., Geologic Features: Death Valley region, California. California Division of Mines and Geology Special Report 106, p. 50.

Wright, L. A., and Prave, A., 1992, Proterozoic-Early Cambrian tectonostratigraphic record of the Death Valley region, California-Nevada, *in* Reed, J., et al., eds. Precambrian rocks of the conterminous United States: Geological Society of America p. 529–533.

Wright, L. A., Thompson, R. A., Troxel, B. W., Pavlis, T. L., DeWitt, E. H., Otton, J. K., Ellis, M. A., Miller, M. G., and Serpa, L. F., 1991, Cenozoic magmatic and tectonic evolution of the east-central Death Valley region, California, *in* Walawender, M. J., and Hanan, Barry B., eds., Geological excursions in southern California and Mexico: Guidebook, Geological Society of America, San Diego, California.

Wright, L.A., Greene, R.C., Cemen, I., Johnson, F.C., and Prave, A.R., 1999, Tectonostratigraphic development of the Miocene-Pliocene Furnace Creek Basin and related features, Death Valley region, California: *in* Wright, L A., and Troxel, B.W. eds., Tertiary basins in the Death Valley region, California: Geological Society of America Special Paper 333, p. 87–114.